Rockets

JOSEPH A. ANGELO, JR.

Facts On File
An imprint of Infobase Publishing

ROCKETS

Facts On File, Inc.
An imprint of Infobase Publishing
132 West 31st Street
New York NY 10001

Library of Congress Cataloging-in-Publication Data
Angelo, Joseph A.
    Rockets / Joseph A. Angelo, Jr.
        p. cm. — (Frontiers in space)
    Includes bibliographical references and index.
    ISBN 0-8160-5771-0
    1. Rockets (Aeronautics) 2. Rocketry. I. Title. II. Series.
    TL782.A48 2006
    621.43'56—dc22   2005016666

Facts On File books are available at special discounts when purchased in bulk quantities for businesses, associations, institutions, or sales promotions. Please call our Special Sales Department in New York at (212) 967-8800 or (800) 322-8755.

You can find Facts On File on the World Wide Web at
http://www.factsonfile.com

Text design by Erika K. Arroyo
Cover design by Salvatore Luongo
Illustrations by Sholto Ainslie

Printed in the United States of America

VB FOF 10 9 8 7 6 5 4 3 2 1

This book is printed on acid-free paper.

To my sister, Mary, and her children:
Kerry, James, and Jean.

Contents

Preface

··

*It is difficult to say what is impossible, for the dream of
yesterday is the hope of today and the reality of tomorrow.*

—Robert Hutchings Goddard

Frontiers in Space is a comprehensive multivolume set that explores the scientific principles, technical applications, and impacts of space technology on modern society. Space technology is a multidisciplinary endeavor, which involves the launch vehicles that harness the principles of rocket propulsion and provide access to outer space, the spacecraft that operate in space or on a variety of interesting new worlds, and many different types of payloads (including human crews) that perform various functions and objectives in support of a wide variety of missions. This set presents the people, events, discoveries, collaborations, and important experiments that made the rocket the enabling technology of the space age. The set also describes how rocket propulsion systems support a variety of fascinating space exploration and application missions—missions that have changed and continue to change the trajectory of human civilization.

The story of space technology is interwoven with the history of astronomy and humankind's interest in flight and space travel. Many ancient peoples developed enduring myths about the curious lights in the night sky. The ancient Greek legend of Icarus and Daedalus, for example, portrays the age-old human desire to fly and to be free from the gravitational bonds of Earth. Since the dawn of civilization, early peoples, including the Babylonians, Mayans, Chinese, and Egyptians, have studied the sky and recorded the motions of the Sun, the Moon, the observable planets, and the so-called fixed stars. Transient celestial phenomena, such as a passing comet, a solar eclipse, or a supernova explosion, would often cause a great deal of social commotion—if not out right panic and fear—because these events were unpredictable, unexplainable, and appeared threatening.

It was the ancient Greeks and their geocentric (Earth-centered) cosmology that had the largest impact on early astronomy and the emergence of Western Civilization. Beginning in about the fourth century B.C.E., Greek philosophers, mathematicians, and astronomers articulated a geocentric model of the universe that placed Earth at its center with everything else revolving about it. This model of cosmology, polished and refined in about 150 C.E. by Ptolemy (the last of the great early Greek astronomers), shaped and molded Western thinking for hundreds of years until displaced in the 16th century by Nicolaus Copernicus and a heliocentric (sun-centered) model of the solar system. In the early 17th century, Galileo Galilei and Johannes Kepler used astronomical observations to validate heliocentric cosmology and, in the process, laid the foundations of the Scientific Revolution. Later that century, the incomparable Sir Isaac Newton completed this revolution when he codified the fundamental principles that explained how objects moved in the "mechanical" universe in his great work *Principia Mathematica*.

The continued growth of science over the 18th and 19th centuries set the stage for the arrival of space technology in the middle of the 20th century. As discussed in this multivolume set, the advent of space technology dramatically altered the course of human history. On the one hand, modern military rockets with their nuclear warheads redefined the nature of strategic warfare. For the first time in history, the human race developed a weapon system with which it could actually commit suicide. On the other hand, modern rockets and space technology allowed scientists to send smart robot exploring machines to all the major planets in the solar system (including tiny Pluto), making those previously distant and unknown worlds almost as familiar as the surface of the Moon. Space technology also supported the greatest technical accomplishment of the human race, the Apollo Project lunar landing missions. Early in the 20th century, the Russian space travel visionary Konstantin E. Tsiolkovsky boldly predicted that humankind would not remain tied to Earth forever. When astronauts Neil Armstrong and Edwin (Buzz) Aldrin stepped on the Moon's surface on July 20, 1969, they left human footprints on another world. After millions of years of patient evolution, intelligent life was able to migrate from one world to another. Was this the first time such an event has happened in the history of the 14-billion-year-old universe? Or, as some exobiologists now suggest, perhaps the spread of intelligent life from one world to another is a rather common occurrence within the galaxy. At present, most scientists are simply not sure. But, space technology is now helping them search for life beyond Earth. Most exciting of all, space technology offers the universe as both a destination and a destiny to the human race.

Each volume within the Frontiers in Space set includes an index, a chronology of notable events, a glossary of significant terms and concepts, a helpful list of Internet resources, and an array of historical and current print sources for further research. Based upon the current principles and standards in teaching mathematics and science, the Frontiers in Space set is essential for young readers who require information on relevant topics in space technology, modern astronomy, and space exploration.

Acknowledgments

I wish to thank the public information specialists at the National Aeronautics and Space Administration (NASA), the National Oceanic and Atmospheric Administration (NOAA), the United States Air Force (USAF), the Department of Defense (DOD), the Department of Energy (DOE), the National Reconnaissance Office (NRO), the European Space Agency (ESA), and the Japanese Aerospace Exploration Agency (JAXA), who generously provided much of the technical materials used in the preparation of this set. The staff at the Evans Library of Florida Tech also provided valuable assistance—as they have for the last three decades. Special acknowledgment is made of the efforts of Frank Darmstadt and other members of the editorial staff at Facts On File, whose diligent attention to detail helped transform an interesting concept into a set of publishable works. The support of two other very special people merits public recognition here. The first individual is my personal physician, Dr. Charles S. Stewart III, M.D., whose medical skills allowed me to successfully complete this volume of Frontiers in Space. The second individual is my wife, Joan, who, as she has done for the past 40 years, provided the loving spiritual and emotional environment so essential in the successful completion of any undertaking in life, including the production of this work.

Introduction

. . . .

Modern launch vehicles are sophisticated propulsion machines that harness the action-reaction principle to provide access to outer space. *Rockets* examines the evolution of the rocket from the fire arrows of ancient China to the incredibly powerful launch vehicles that enable space travel and allow us to meet the universe face to face. Emerging out of World War II and then greatly improved during the cold war, the modern ballistic missile dramatically changed the strategic equation for warfare and military conflict. Then, the ballistic missile was converted for use in space launches and brought about the arrival of the space age—forever changing the destiny of the human race.

Rockets describes the historic events, scientific principles, and technical breakthroughs that now allow complex launch vehicles to send satellites into orbit around Earth or sophisticated robot spacecraft to mysterious worlds in our solar system. A generous number of sidebars are strategically positioned throughout the book to provide expanded discussions of fundamental scientific concepts and rocket engineering techniques. There are also capsule biographies of important rocket scientists and aerospace engineers to let the reader appreciate the human dimension in rocketry.

It is especially important to recognize that the modern rocket is now the enabling technology for many exciting future options for the human race. Awareness of these pathways should prove inspiring to those students now in high school and college who will become the scientists, engineers, and astronauts of tomorrow. Why are such career choices important? Future advances in rocket propulsion for space travel no longer represents a simple option of governments that can be pursued or not, depending on political circumstances. Rather, continued advances in rocketry represent a technical, social, and psychological imperative for the human race. We can decide to become a spacefaring species as part of our overall sense of being and purpose, or we can ignore the challenge and opportunity before us, and turn our collective backs on the universe. The latter choice would confine future generations to life on just one planet around an average star in the outer regions of the Milky Way Galaxy. The former choice makes the human race a spacefaring species with all the exciting social and technical impacts that decision entails.

Ever mindful of the impact of science and technology on society, this book examines the role the modern rocket has played in human development since the middle of the 20th century and then projects the expanded role the rocket can play in this century and beyond. Routine access to space now supports such important areas as global security and defense, a better understanding of Earth as a complex environmental system, weather forecasting, natural hazard warning, global communications, and navigation. Orbiting scientific instruments provide us a richer understanding of how the universe works. Because of the rocket, scientists are now searching for life on other worlds in the solar system. Later in this century, they may even be able to answer the age-old philosophical question: Are we alone in this vast universe?

Rockets also shows that the development of modern rocket technology did not occur without problems, issues, and major financial commitments. Selected sidebars within the book address some of the most pressing contemporary issues associated with the application modern rocket technology. These include the global proliferation of military missile technology, the possible extension of terrestrial conflicts to outer space, the growing problem of space debris, and the risk human crews face during rocket launches and space travel. *Rockets* also describes how future advances in rocket technology will exert interesting social, political, and technical influences that should extend well beyond this century. Some of these potential impacts include permanent human settlements on the Moon and Mars, very smart robot explorers traveling beyond the outermost regions of the solar system, the operation of a planetary asteroid defense system, the discovery of life (extinct or existing) beyond Earth, and the emergence of a solar-system civilization. Advanced rocket propulsion systems are the underlying and enabling technology for such interesting future developments.

Rockets has been carefully crafted to help any student or teacher who has an interest in rocketry discover what rockets are, where they came from, how they work, and why they are so important. Although the international (or SI) unit system is the preferred "language" of modern science and engineering, *Rockets* also provides units in terms of the traditional or American engineering system of units. For example, masses are given in both pounds-mass (lbm) and kilograms (kg). This editorial approach should help any student or teacher better appreciate science and engineering in a global context. The reader can also more easily bridge the gap between commonly encountered American units to possibly less familiar—although very important—metric units. The back matter contains a chronology, glossary, and an array of historical and current sources for further research. These sections should prove especially helpful for readers who need additional information on specific terms, topics, and events in rocketry.

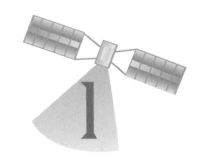

From Fire Arrows to Mars Missions

Rocketry is the art of making rockets and embodies that branch of engineering science that deals with rockets—including theory, research, development, experimentation, and application. This chapter explores the evolution of the rocket from primitive gunpowder devices used to startle enemy troops in ancient China to sophisticated reaction engines capable of placing human beings on the Moon and thrusting scientific instruments to other worlds throughout the solar system and beyond.

✧ Gunpowder Rockets

According to certain historical records, the Chinese were the first to use gunpowder rockets, which they called fire arrows, in military applications. In the battle of Kaifung-fu (1232 C.E.), for example, fire arrows helped the Chinese repel Mongol invaders.

The fire arrow was an early gunpowder rocket attached to a large bamboo stick. The Chinese developed this device about 1,000 years ago to confuse and startle enemy troops. Exactly when the idea of the rocket first emerged in ancient China is not clear. Some science historians suggest that Chinese chemists had formulated gunpowder by the first century C.E. Gunpowder is an explosive mixture of potassium nitrate (KNO_3), sulfur, and charcoal (powdered carbon). Because this mixture is generally dark gray or black, the term *black powder rocket* also appears in the literature of rocketry. The early Chinese were initially content to simply use this explosive chemical mixture to make fireworks for festivals. They would stuff gunpowder into hollowed pieces of bamboo and then toss the devices into a fire where they soon exploded, producing the desired festive effect.

Quite possibly one of these firecracker devices did not explode, but rather shot out of the fire like a rocket—in response to the propulsive action-reaction principle caused by the escaping hot gases from burning

A 13th-century Chinese warrior launches a fire arrow. *(Drawing courtesy of NASA)*

gunpowder. (The physical principles of rocket propulsion are discussed in chapter 2.) At that point, an inquisitive Chinese "rocketeer" probably got the clever idea of attaching one of these bamboo tubes filled with gunpowder to an arrow or long stick. The famous Chinese fire arrow was born.

Whatever the actual creative pathway of the rocket's discovery in ancient China, the Battle of Kaifung-fu in 1232 represents the first reported use of a gunpowder-fueled rocket in warfare. During this battle, Chinese troops used a barrage of rocket-propelled fire arrows to startle and defeat a band of invading Mongolian warriors. In an early attempt at passive guided missile control, Chinese rocketeers attached a long stick to the end of the fire arrow rocket. The long stick kept the center of pressure behind the rocket's center of mass during flight. Although the addition of this long stick helped somewhat, the flight of the rocket-propelled fire arrows still remained quite erratic and highly inaccurate. The heavy stick also reduced the range of these early gunpowder-fueled rockets. Despite the limitations of the fire arrow, the invading Mongol warriors quickly learned from their unpleasant experience at the Battle of Kaifung-fu and soon adopted the interesting new weapon for their own use. As a result, nomadic Mongol warriors spread rocket technology westward when they invaded portions of India, the Middle East, and Europe.

Before this book describes the migration of the gunpowder rocket to other parts of the world, there is another interesting rocket story from China that deserves mention here. Rocketry legend suggests that around 1500, a lesser-known Chinese official named Wan-Hu conceived of the idea of flying through the air in a rocket-propelled chair. He ordered the construction of a chair-kite structure to which were attached 47 fire arrow rockets. Then, serving as his own test pilot, Wan-Hu bravely sat in the chair and ordered his servants to simultaneously light the fuses to all the rockets. Forty-seven servants, each carrying a small torch, rushed forward in response to their master's command. Dutifully, they lit the fuses and then dashed back to safety. Suddenly, there was a bright flash and a tremendous roar. The air was filled with billowing clouds of gray smoke. Unfortunately, Wan-Hu and his rocket-propelled chair vanished in the explosion—perhaps reaching the heavens more suddenly than he intended. While science historians regard this story as more legend than fact, it represents the first reported attempt to use the rocket as a means of transportation. Previous applications of the gunpowder rocket were related to either warfare or fireworks for festivals.

Mongol warriors quickly incorporated the Chinese fire arrow into their arsenal and employed these rockets in various battles during invasions of India, the Middle East, and portions of Europe in the 13th and 14th century. Their actions are considered the primary cause for the spread of rocket technology out of China. However, the far eastern travels of the legendary Italian adventurer Marco Polo (ca.1254–ca.1324) may have also contributed to the arrival of gunpowder and the rocket technology in medieval Europe. Some science historians further suggest that a British Franciscan monk, named Roger Bacon (ca.1214–92) independently developed a formula for gunpowder in about 1242 and then improved the burn rate of his particular mixture by using distilled potassium nitrate.

Whatever the case, by the 14th century, gunpowder-rocket technology migrated to western Europe, where over the next few centuries military engineers attempted to develop and expand its role in warfare. As a result, the gunpowder rocket ended up in many European Renaissance arsenals as a primitive bombardment weapon. The first recorded military use of the gunpowder rocket in western Europe occurred in 1379, during the siege of Chioggia, Italy. About four decades later, in 1420, the Italian military engineer Joanes de Fontana wrote *Bellicorum Instrumentorum Liber* (Book of war machines). His speculative work suggested several military applications of gunpowder rockets, including a rocket-propelled battering ram and a rocket-propelled torpedo. In 1429, the French army used gunpowder rockets to defend the city of Orléans.

For the next two centuries, military engineers throughout Europe continued to test various types of gunpowder rockets as an alternative to early cannons. By the end of the Renaissance, most European armories bristled with a collection of gunpowder rockets—all technical descendants of the Chinese fire arrow. However, artillery improvements eventually made the cannon more effective in battle, and military rockets shifted into the background.

While the use of the gunpowder rocket as a weapon declined in 16th- and 17th-century Europe due to improvements in cannon technology, the application of the rocket in fireworks (that is, festive devices called sky rockets) continued to evolve into a highly refined technical art. In most European countries, the gunpowder rocket flourished as the centerpiece of the fireworks displays that amused royalty on festive occasions. One enterprising 16th-century fireworks maker in Germany, Johann Schmidlap, wanted to make his displays go higher into the air, so he invented what rocket engineers today call the step-rocket. Schmidlap used a large skyrocket (the first stage) to carry a smaller skyrocket (the second stage) as its payload. When the large rocket burned out, the smaller one continued to climb to a higher altitude before exploding and showering the night sky with colorful glowing cinders. Of course, he had to do a great deal of pyrotechnic tinkering to get the second stage to ignite at just the right moment. Although the physics of rocket propulsion is now much better understood, the manufacture of sophisticated modern skyrockets and pyrotechnic displays is still very much a technical art, which traces its heritage back to European fireworks makers like Johann Schmidlap. It should be noted that aerospace historians generally attribute the concept of the modern multistage rocket to the Russian space travel pioneer Konstantin Tsiolkovsky. (His important contributions to rocketry are discussed shortly.)

After emerging from China, the rocket also found military application in the Middle East and in India. For example, in 1249, Arab troops used rockets in their unsuccessful attempt to thwart invading Crusader troops under Louis IX of France, during the siege of Damietta—a strategic medieval port city in northern Egypt. Between 1280 and 1290, the Arab military historian, Al-Hasan al Rammah, wrote *The Book of Fighting on Horseback and War Strategies*. In this work, he mentioned Chinese fire arrows and provided instructions for making both gunpowder and rockets.

During the 16th and 17th centuries, Europe experienced a period of profound changes in intellectual thought—often referred to as the Scientific Revolution. Nicolaus Copernicus (1473–1543) began this process by causing a revolution in astronomy with his deathbed publication of *On the Revolutions of Celestial Spheres*. The book flew in the face of almost two millennia of Aristotle's geocentric astronomy and endorsed a heliocentric model of the universe in which Earth, like the other known planets, revolved around the Sun. In the early part of the 17th century, the telescopic observations of Galileo Galilei (1564–1642) and the laws of planetary motion developed by Johannes Kepler (1571–1630) reinforced the Copernican revolution. Because of his meticulous experiments and careful attention to physical observations, Galileo is often regarded as the first modern scientist. In the late 17th century, Sir Isaac Newton (1642–1727) capped the scientific revolution by developing and publishing his three laws of motion and the universal law of gravitation. These important scientific principles allowed scientists to explain in precise mathematical terms the motion of almost every object observed in the universe, from an apple falling to the ground, to the trajectory of projectiles fired from a cannon, to planets orbiting the Sun.

Stimulated by the scientific revolution, 17th-century military engineers and strategists began to revisit the role of the rocket in warfare. Both the Dutch and the Germans conducted experiments with military rockets, and, in 1680, Russian czar Peter the Great established a military rocket–manufacturing facility near Moscow. Despite such activities, the military use of the rocket in western Europe remained in the background. European armies elected to use the improved gunpowder rockets primarily as signaling devices and to illuminate battlefields during nighttime engagements. However, elaborate fireworks displays still remained a favorite pastime for European royalty. In contrast, during the late 18th century, the gunpowder rocket served as an important weapon in South Asia, where Indian rulers successfully used them in several battles against British army units.

In the late 1700s, Rajah Hyder Ali—prince of Mysore, India—used iron-case stick rockets to defeat British troops. Profiting from their adverse rocket experience in India, British military engineers, led by Sir William Congreve, improved the design of captured Indian rockets and developed a series of more efficient bombardment rockets. Congreve's assault rockets ranged in mass from about 8 to 136 kilograms and consisted of two basic types: the shrapnel (case-shot) rocket and the incendiary rocket. In the early 19th century, British troops would often use the shrapnel rocket as a substitute for artillery. As this gunpowder rocket flew over enemy soldiers, its exploding warhead showered the battlefield with rifle balls and sharp pieces of metal. The British would use the incendiary rocket during the siege of a city or during a naval engagement. The warhead of the incendiary rocket was filled with sticky, flammable material that quickly started fires when it impacted in an enemy city or in the rigging of an enemy sailing ship.

The British used both types of military rocket effectively during the Napoleonic Wars and the War of 1812. Perhaps the most famous application of Congreve's rockets was the British bombardment of the American Fort McHenry, which occurred in the War of 1812. This rocket attack is now immortalized in the "rockets' red glare" phrase heard in "The Star-Spangled Banner."

The British inventor William Hale (1797–1870) attempted to improve the inherently low accuracy of Congreve's stick-guided rockets through a technique called spin stabilization. He eliminated the bulky and cumbersome wooden guide sticks that Congreve's rockets used for stability and guidance. Instead, Hale made his rockets stable during flight through the invention of a special nozzle configuration, which vectored some of the escaping propellant gases to spin or rotate the rocket about its longitudinal axis. Soon after he patented his spinning rocket in 1844, Hale sold the manufacturing rights to these devices to the United States government. As a result of his entrepreneurial actions, U.S. Army units used Hale's bombardment rockets during the Mexican-American War (1846–48). In fact, the first battlefield use of a spin-stabilized rocket took place during the siege of Veracruz, Mexico, in March 1847. Hale's rockets were also used during the U.S. Civil War and achieved mixed results.

The British Army experimented with Hale's new rockets during the Crimean War (1853–56) and eventually adopted them in about 1867

Sir William Congreve
(1772–1828)

While a colonel of artillery, the British military engineer Sir William Congreve examined black powder (gunpowder) rockets captured during late 18th century battles in India. He then supervised the development of a series of improved British military rockets. In 1804, Congreve wrote *A Concise Account of the Origin and Progress of the Rocket System*. During this period, he also supervised the construction of a wide variety of gunpowder-fueled military rockets. His rockets ranged in mass from about 330 pounds (150 kg) down to 18 pounds (8 kg).

Congreve's efforts provided the British army with two basic types of assault rockets: the shrapnel (case-shot) rocket and the incendiary rocket. The shrapnel rocket often substituted for artillery. When this weapon flew over enemy troops, its exploding warhead showered the battlefield with rifle balls and pieces of sharp metal. Congreve filled the warhead of his incendiary rocket with sticky, flammable materials that quickly started fires when it impacted in an enemy city or in the rigging of an enemy sailing ship. His pioneering work on these early solid propellant rockets represents an important technical step in the overall evolution of the modern military rocket.

British forces used Congreve's rockets quite effectively in large-scale bombardments during the Napoleonic Wars and the War of 1812. Perhaps the most famous application of Congreve's rockets took place in August 1814 during the British bombardment of the American Fort McHenry in the War of 1812. Throughout this attack, a young American lawyer and prisoner-exchange negotiator, Francis Scott Key (1780–1843), remained under guard, first onboard the British ship, HMS *Surprise*, later on a sloop anchored behind the British battle fleet. Throughout the night, Key witnessed the relentless naval bombardment of the fort, including the glowing red flames from Congreve's rockets, as they wobbled skyward towards the fort from numerous small launching boats. At dawn, the American flag still flew over the fort despite the massive British bombardment. Key immortalized the event and Congreve's rockets, when he wrote of the "rocket's red glare" in a special poem that became "The Star-Spangled Banner."

—primarily for use in military engagements in colonies in Africa and Asia. British soldiers nicknamed one of the most popular versions of the Hale military rocket the "24-pounder." This 24-pound (11-kg) spin-stabilized gunpowder rocket had an explosive warhead, was easily mule-packed into the field, and had a combat range of between 0.6 mile (1 km) and 2.5 miles (4 km)—depending on firing angle, environmental conditions, and geographic factors. However, improvements in artillery technology again outpaced developments in military rocketry; by the beginning of the 20th century, rockets once again became more a matter of polite speculation, rather than a widely accepted military technology. For example, by 1919, the

British Royal Army officially declared the Hale rocket obsolete—although by the 1890s the 24-pounder and other Hale rockets had already been overtaken in range, accuracy, and safety by conventional artillery.

But World War II stimulated another round in the centuries-old competition between the military rocket and conventional artillery. By the middle portion of the 20th century the technology of military rockets had significantly advanced, and the solid-propellant rocket again became an integral part of the modern battlefield. This time, however, the technical descendants of the ancient Chinese fire arrows were also joined by a much more powerful companion—the liquid-propellant rocket, whose arrival forever changed the nature of strategic warfare and opened up the heavens to exploration by human beings and smart machines. The creation and use of the nuclear weapon at the end of World War II was another technological development that significantly altered the rocket versus cannon competition. Improvements in nuclear weapon design soon provided rocket engineers with a payload of unprecedented destructiveness. During the superpower arms race of the cold war, military strategists in both the United States and the former Soviet Union quickly embraced the nuclear-

Resembling an ancient fire arrow, this U.S. Army Chaparral missile streaks toward its test target at the White Sands Missile Range (1977). The heat-seeking Chaparral missile (now retired) was 2.9 meters long, 0.13 meter in diameter and had a total mass (including solid propellant) of 84 kilometers. Designed to protect 20th-century soldiers and their equipment from attack by aircraft, the Chaparral missile had an operational range of more than one kilometer. *(U.S. Army/White Sands Missile Range)*

armed ballistic missile as the ultimate weapon—forever changing trajectory of human history.

✧ Dreams of Space Travel and the Birth of Astronautics

Starting in the mid-19th century, the French writer and technical visionary Jules Verne (1828–1905) created modern science fiction and along with it the dream of space travel. Perhaps Verne's greatest influence on the development of space travel was his 1865 novel, *De la terre à la lune* (*From the Earth to the Moon*). In this fictional work, Verne gave his readers a somewhat credible account of a human voyage to the Moon. Verne's travelers are blasted on a journey around the Moon in a special hollowed-out capsule that is fired from a very large cannon. The writer correctly located the cannon at a low-latitude site in Florida. Of course, scientists recognized that the acceleration of Verne's proposed capsule down the barrel of this huge cannon would have immediately crushed the three intrepid space travelers inside (as if that were not bad enough, the capsule itself would have burned up traveling at escape velocity speed through Earth's atmosphere). Despite its obvious technical limitations, this tale made spaceflight appear possible for the first time in history.

Although Verne did not properly connect the rocket as the enabling technology for space travel, his famous story did correctly prophesize the use of small reaction rockets to control the attitude of the ballistic capsule during its flight through space. Verne was not a scientist or engineer, but his literary skills served as an important source of inspiration for those scientists and engineers who actually responded to the challenge of interplanetary space travel. In particular, the three great pioneers of astronautics—Konstantin Tsiolkovsky, Robert Goddard, and Hermann Oberth—would soon independently make the important and necessary connection between powerful liquid-propellant rockets and space travel. Each of these rocket pioneers also personally acknowledged the works of Jules Verne as a key childhood stimulus in developing their lifelong interest in space travel. The great French novelist, who died in Amiens, France, on March 24, 1905, not only wrote delightful stories that pleased millions of readers, he lit the flame of imagination for those who would actually create the modern rockets needed to free humankind from the bonds of Earth.

The 19th-century technofiction works of Jules Verne provided youthful inspiration for many of the rocket pioneers, who made space travel a reality in the 20th century. This 1955 postage stamp from Monaco honors the French writer and his famous tale *From the Earth to the Moon*. As if to herald the arrival of the space age (in 1957), a stylized winged rocket also appears in the beautifully engraved stamp's design. *(Courtesy of the author)*

Because of Jules Verne, rocket-propulsion based space travel became first the technical dream, and then, the technical reality, of the 20th century.

The first space-travel visionary we discuss here is the Russian schoolteacher Konstantin Tsiolkovsky. He wrote a series of articles and books about the theory of rocketry and space flight at the turn of the century. Among other things, his pioneering works suggested the necessity for liquid-propellant rockets—the very devices that the American physicist Robert Goddard would soon develop. Because of the geopolitical circumstances in czarist Russia, Goddard and many other scientists outside of Russia were unaware of Tsiolkovsky's work. Today, Tsiolkovsky is regarded

JULES VERNE'S GIANT CANNON

Scientists use the following equation to make a rough calculation of the acceleration (a) a theoretical projectile-like space capsule would need to reach the Moon when shot from Earth:

$$v^2 = v_0 + 2ax$$

First, assume that the initial velocity (v_0) of the space capsule is zero when it sits at the bottom end of the giant cannon. Next, assume that the capsule's final velocity (v) as it leaves the cannon's barrel is equal to the escape velocity (v_{esc}). To keep matters simple, scientists often initially neglect both friction down the cannon barrel and aerodynamic friction when the capsule flies out of the cannon at great speed into the atmosphere. Of course, neither these are really factors to ignore, but, for the moment, these assumptions help keep the analysis as simple and uncluttered as possible. Next, assume that the giant cannon is 0.62 mile (1 km) long. This gives x in the above equation a value of 3,280 feet (1,000 m). Scientists also know that the escape velocity from Earth is about 36,736 feet per second (11,200 m/s).

Now rearrange the first equation into a more useful form and write:

$$a = v^2/(2x)$$

Substituting the values for v_{esc} and x into this equation, the result is:

$$a = (36,736 \text{ ft/s})^2 / [(2) (3,280 \text{ ft})] = 205,720 \text{ ft/s}^2 (62,720 \text{ m/s}^2)$$

The units may look a little unusual, but they are perfectly correct for acceleration. Scientists know that a mass at the surface of Earth experiences acceleration due to gravity (g) of about 32.2 feet per second squared (9.8 m/s²). In sharp contrast, a 0.62-mile (1-km) long version of Jules Verne's Moon cannon would shoot the capsule out at 6,400 gs. This is an incredible acceleration. Any person inside such a capsule would experience 6,400 times their normal weight and would be squashed into an unrecognizable puddle of mush at the bottom of the accelerating projectile. As a point of comparison, astronauts and cosmonauts experience brief periods of acceleration ranging from about 3 g to 5 g, when they ride onboard different rocket vehicles into space.

KONSTANTIN EDUARDOVICH TSIOLKOVSKY
(1857–1935)

The Russian schoolteacher Konstantin Eduardovich Tsiolkovsky is one of the three founding fathers of astronautics. At the beginning of the 20th century, Tsiolkovsky worked independent of Robert Goddard and Hermann Oberth, but the three shared and promoted the important common vision of using rockets for interplanetary travel.

Tsiolkovsky, a nearly deaf Russian schoolteacher, was a theoretical rocket expert and space-travel pioneer light-years ahead of his time. This brilliant schoolteacher lived a simple life in isolated, rural towns within czarist Russia. Yet, despite his isolation from the mainstream of scientific activity, he somehow wrote with such uncanny accuracy about modern rockets and space travel that he cofounded the field of astronautics, the science of space flight. Primarily a theorist, he never constructed any of the rockets he proposed in his incredibly prophetic books.

His 1895 book, *Dreams of Earth and Sky*, included the concept of an artificial satellite orbiting Earth. Many of the most important principles of astronautics appeared in his seminal 1903 work, *Exploration of Space by Reactive Devices*. This book linked the use of the rocket to space travel—especially the high-performance liquid hydrogen and liquid oxygen rocket engine. Tsiolkovsky's 1924 work *Cosmic Rocket Trains* introduced the concept of the modern multistage rocket. His books inspired many future Russian cosmonauts, space scientists, and rocket engineers, including Sergei Korolev, whose powerful rockets helped fulfill Tsiolkovsky's predictions.

At 16, the young Tsiolkovsky ventured to Moscow, where he studied mathematics, astronomy, mechanics, and physics. Almost completely deaf since childhood, he used an ear trumpet to listen to lectures and struggled with a meager weekly food allowance of just a few kopecks (pennies). Three years later Tsiolkovsky completed his studies and returned home. After passing the schoolteacher's examination, he began his teaching career at a rural school in Borovsk, located about 62 miles (100 km) from Moscow. In Borovsk, he met and married Varvara Sokolovaya. For more than a decade, he remained a young provincial schoolteacher in Borovsk. Then, in 1892, Tsiolkovsky moved to another teaching post in Kaluga, where he remained until he retired in 1920.

As he began his teaching career in rural Russia, Tsiolkovsky also turned his fertile mind to science, especially concepts about rockets and space travel. Despite his severe hearing impairment, Tsiolkovsky's tenacity and self-reliance allowed him to become an effective teacher and also to make significant contributions to the fields of aeronautics and astronautics.

While teaching in Borovsk, Tsiolkovsky used his own meager funds to construct the first wind tunnel in Russia. He did this so he could experiment with airflow over various streamlined bodies. He also began making models of gas-filled, metal-skinned dirigibles. His interest in aeronautics served as a stimulus for his more visionary work involving the theory of rockets and their role in space travel. As early as 1883, he accurately described the weightlessness conditions of space in an article entitled "Free Space." By 1898, he correctly linked the rocket to space travel and concluded that the rocket would have to be a liquid-fueled chemical rocket in order to achieve the necessary escape velocity—to completely escape from Earth's gravity a spacecraft would need to reach a minimum velocity of approximately seven miles per second (11 km/s).

Many of the fundamental principles of astronautics were described in his seminal work, *Explo-*

ration of Space by Reactive Devices. This important theoretical treatise showed that space travel was possible using the rocket. Another pioneering concept found in the book is a design for a liquid-propellant rocket that used liquid hydrogen and liquid oxygen. Tsiolkovsky delayed publishing the important document until 1903. One possible reason for the delay was the death of his son, Iganty, who committed suicide in 1902.

Because Tsiolkovsky was a village teacher in an isolated, rural area of prerevolutionary Russia, his pioneering work in aeronautics and astronautics went essentially unnoticed by the world scientific community. In those days, few people in czarist Russia cared about space travel, so he never received significant government funding to pursue any type of practical demonstration of his innovative concepts. His visionary ideas included the spacesuit, space stations, multistage rockets, large habitats in space, the use of solar energy, and closed life-support systems.

Following the Russian Revolution of 1917, the new Soviet government grew interested in rocketry and rediscovered Tsiolkovsky's amazing work. The Soviet government honored him for his previous achievements in aeronautics and astronautics and encouraged him to continue his pioneering research. He received membership in the Soviet Academy of Sciences in 1919 and the government granted him a pension for life in 1921 in recognition of his overall teaching and scientific contributions.

Tsiolkovsky used the free time of retirement to continue to make significant contributions to astronautics. In his 1924 book, *Cosmic Rocket Trains,* he recognized that on its own a single stage rocket would not be powerful enough to escape Earth's gravity and promoted the concept of a staged rocket, which he called a rocket train. He died in Kaluga on September 19, 1935. His epitaph conveys the important message: *"Mankind will not remain tied to the Earth forever."*

Tsiolkovsky's visionary writings provided career inspiration for many future Russian aerospace engineers, including Sergei Korolev. As part of the former Soviet Union's celebration of the centennial of Tsiolkovsky's birth, Korolev received permission to use a powerful Russian military rocket as a space lift vehicle to launch *Sputnik 1*—the world's first artificial satellite on October 4, 1957. This technical achievement is generally regarded as the birth of the space age.

At the beginning of the 20th century, Konstantin E. Tsiolkovsky helped to establish the field of astronautics and in the process also became the Father of Russian Rocketry. Tsiolkovsky's portrait appears on this 1964 postage stamp—issued by the former Soviet Union to honor his great contributions to rocketry and space travel. *(Courtesy of the author)*

Dr. Robert H. Goddard—the reclusive genius who is widely acknowledged as the "Father of American Rocketry"—is shown with a steel combustion chamber and rocket nozzle in this 1915 photograph. *(NASA)*

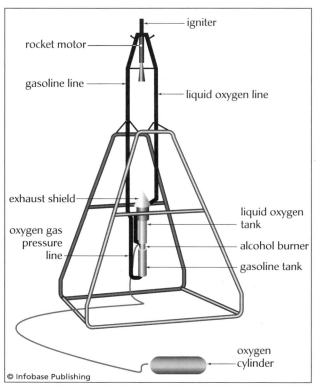

A schematic diagram of the world's first liquid-propellant rocket, constructed by Dr. Robert H. Goddard and flown on March 16, 1926

as the "Father of Russian Rocketry" and one of three cofounders of astronautics—the science of space flight. His tombstone bears the prophetic inscription "Mankind will not remain tied to the Earth forever!"

Similarly, the brilliant physicist Robert Hutchings Goddard (1882–1945) is regarded as the "Father of American Rocketry" and the developer of the practical modern rocket. Goddard started working on the physical principles underlying rockets propulsion about 1912 and by 1915 began testing solid-fuel models. In 1917, after the United States entered World War I, Goddard worked to perfect rockets as weapons. One of his designs became the technical forerunner of the bazooka—a tube-launched anti-tank rocket. Goddard's device—about 18 inches (45.7 cm) long and one inch (2.54 cm) in diameter—was tested in 1918, but the war ended before it could be used against enemy tanks.

In 1919, Goddard published an important technical paper, "A Method of Reaching Extreme Altitudes," in which he concluded that the rocket actually would work better in the vacuum of outer space than in Earth's atmosphere. At the time, Goddard's "radical" (but correct) suggestion cut sharply against the popular (but incorrect) belief that a rocket needed air to "push against." He also suggested that a multistage rocket could reach very high altitude and attain sufficient velocity to "escape from Earth." Unfortunately, the press scoffed at his ideas and the general public failed to appreciate the great scientific merit of this paper.

Despite the adverse publicity, Goddard continued to experiment with rockets, but now he intentionally avoided any interactions with members

Dr. Robert H. Goddard and the liquid oxygen-gasoline rocket in the launch frame from which he fired it on March 16, 1926, in a snow-covered field at Auburn, Massachusetts. The event represents the birth of the liquid-propellant rocket. *(NASA)*

of the news media. On March 16, 1926, he launched the world's first liquid-fueled rocket from a snow-covered field at his Aunt Effie Goddard's farm in Auburn, Massachusetts. This simple gasoline-and-liquid-oxygen-fueled device rose to an estimated height of just 39 feet (12 m) and landed about 184 feet (56 m) away in a frozen field. Regardless of its initial altitude or range, Goddard's liquid-propellant rocket flew successfully, and the world would never be quite the same. The numerous technical progeny of this simple liquid-propellant rocket have taken human beings into orbit around Earth and to the surface of the Moon. Powerful liquid-propellant rockets have also sent sophisticated space probes on incredible journeys of exploration throughout the solar system and beyond.

After this initial success, Goddard flew other rockets in rural Massachusetts—at least until they started crashing into neighbors' pastures. After the local fire marshal declared that his rockets were a fire hazard, Goddard terminated his New England test program. The famous aviator Charles Lindbergh came to Goddard's rescue and helped him receive a grant from the Guggenheim Foundation.

With this grant, Goddard moved to sparsely populated Roswell, New Mexico, where he could experiment without disturbing anyone. At his Roswell test complex, Goddard developed the first gyro-controlled rocket guidance system. He also flew rockets faster than the speed of sound and at altitudes up to 7,540 feet (2,300 m). Yet, despite his numerous technical accomplishments in rocketry, the U.S. government never really developed an interest in his work. In fact, only during World War II did he receive any government funding; and that was for him to design small rockets to help aircraft take off from navy carriers. By the time he died in 1945, Goddard held more than 200 patents in rocketry. Aerospace engineers and rocket scientists now find it essentially impossible to design, construct, or launch a modern liquid-propellant rocket without using some idea or device that originated from Goddard's pioneering work in rocketry.

✧ The World's First Modern Ballistic Missile

The third cofounder of astronautics was Hermann Julius Oberth, whose writings and leadership promoted interest in rocketry in Germany following World War I. While Goddard worked essentially unnoticed in the United States, a parallel group of "rocketeers" thrived in Germany, centered originally within the German Rocket Society. In 1923, Oberth published a highly prophetic book, *The Rocket into Interplanetary Space.* He proved in this important work that flight beyond the atmosphere was possible. One of the many readers inspired by this book was a brilliant

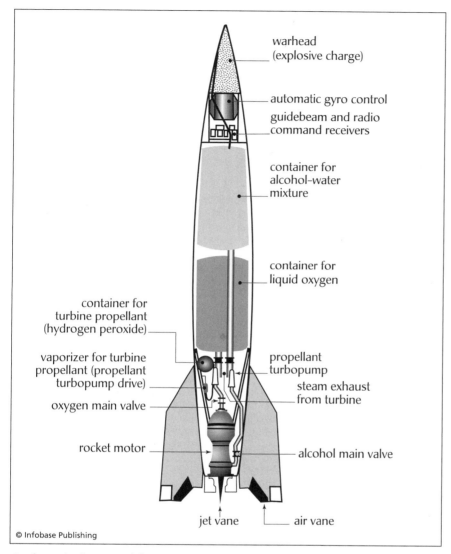

warhead
(explosive charge)

automatic gyro control

guidebeam and radio
command receivers

container for
alcohol-water
mixture

container for
liquid oxygen

container for
turbine propellant
(hydrogen peroxide)

vaporizer for turbine
propellant (propellant
turbopump drive)

propellant
turbopump

steam exhaust
from turbine

oxygen main valve

rocket motor

alcohol main valve

jet vane   air vane

© Infobase Publishing

A schematic diagram of the German V–2 (A-4) rocket—the world's first modern
ballistic missile

young teenager named Wernher von Braun. In 1929, Oberth published
another important book *The Road to Space Travel*. Within this work, he
proposed liquid-propellant rockets, multistage rockets, space navigation,
and guided reentry systems.

From 1939 to 1945, Oberth, along with other German rocket scientists (including von Braun), worked in the military rocket program.

HERMANN JULIUS OBERTH
(1894–1989)

Transylvanian-born German physicist Hermann Julius Oberth was a theoretician who helped establish the field of astronautics early in the 20th century. He originally worked independently of Konstantin Tsiolkovsky and Robert Goddard in his advocacy of rockets for space travel, but later discovered and acknowledged their prior efforts. Unlike Tsiolkovsky and Goddard, Oberth made human beings an integral part of the space travel vision.

His inspirational 1923 publication *The Rocket into Interplanetary Space* provided a comprehensive discussion of all the major aspects of space travel, and his 1929 award-winning book *Roads to Space Travel* popularized the concept of space travel for technical and nontechnical readers alike. Although he remained a theorist and was generally uncomfortable with actually "bending metal," his technical publications and inspiring lectures exerted a tremendous career influence on many young Germans, including the legendary Wernher von Braun.

Oberth was born on June 25, 1894 in the town of Hermannstadt in a German-enclave within the Transylvanian region of Romania (then part of the Austro-Hungarian Empire). At 11, he discovered the works of Jules Verne, especially *From the Earth to the Moon*. Oberth read this novel many times and then, while remaining excited about space travel, questioned the story's technical efficacy. He soon discovered that the acceleration down the barrel of this huge cannon would have crushed Verne's three intrepid explorers and that the capsule itself would have burned up in Earth's atmosphere. But Verne's story made space travel appear technically possible and this important idea thoroughly intrigued Oberth. So, after identifying the technical limitations in Verne's fictional approach, he started searching for a more practical way to travel into space. That search quickly led him to the rocket.

During World War I, Oberth tried to interest the Imperial German War Ministry in developing a long-range military rocket. In 1917, he submitted his specific proposal for a large liquid-fueled rocket. He received a very abrupt response for his efforts. Prussian armaments "experts" within the ministry quickly rejected his plan and reminded him that their experience clearly showed that military rockets could not fly farther than 4.3 miles (7.0 km). Of course, these officials based their negative comments on their own limited experience with inefficient contemporary military rockets, which used gunpowder for propellant. They totally missed Oberth's breakthrough idea involving a better-performing liquid-fueled rocket. Undaunted, Oberth continued to investigate the theoretical problems of rocketry.

In 1918, he married Mathilde Hummel and a year later went back to the University of Munich to study physics. Next, he briefly studied at the University of Göttingen (1920–1921), followed by the University of Heidelberg (1920–1921), before becoming certified as a secondary school mathematics and physics teacher in 1923. At this point in his life, he was unaware at the time of the contemporary rocket theory work of Konstantin Tsiolkovsky in Russia and Robert H. Goddard in the United States. In 1922, he presented a doctoral dissertation on rocketry to the faculty at the University of Heidelberg. Unfortunately, the university committee rejected his dissertation.

Still inspired by space travel, he revised this work and published it in 1923 as *Die Rakete zu den Planetenräumen* (*The Rocket into Interplanetary Space*). This slender book provided a thorough discussion of the major aspects of space travel and its contents inspired many young German scientists and engineers to explore rocketry. Oberth worked as a teacher and writer in the 1920s.

By the mid-1920s, Oberth had discovered and acknowledged the rocketry work of Goddard and Tsiolkovsky. At this point, he also became the organizing principle around which the practical application of rocketry developed in Germany. Oberth served as a leading member of *Verein für Raumschiffahrt* (VfR), the German Society for Space Travel. Members of this technical society conducted critical experiments in rocketry in the late 1920s and early 1930s, until the German army absorbed their efforts and established a large military rocket program.

Oberth was a much better theorist and technical writer than nuts-and-bolts rocket engineer. In 1929, he expanded his ideas concerning the rocket for space travel and human space flight in the award-winning book *Wege zur Raumschiffahrt* (*Roads to Space Travel*). His book helped popularize the concept of space travel for both technical and nontechnical audiences. As newly elected president of the VfR, Oberth used some of the book's prize money to fund rocket engine research within the society. Young engineers like Wernher von Braun had a chance to experiment with liquid-propellant engines, including one of Oberth's own concepts, the *Kegeldüse* (conic) engine design. In this visionary book, Oberth also anticipated the development of electric rockets and ion propulsion systems.

Throughout the 1930s, Oberth continued to work on liquid-propellant rocket concepts and on the idea of human spaceflight. In 1938, Oberth joined the faculty at the Technical University of Vienna. There, he participated briefly in a rocket-related project for the German Air Force. In 1940, he became a German citizen. The following year, he joined von Braun's rocket development team at Peenemünde.

But Oberth only worked briefly with von Braun's military rocket team at Peenemünde and, in 1943, transferred to another location to work on solid-propellant antiaircraft rockets. At the end of the war, Allied forces captured him and placed him in an internment camp. Upon release, he left a devastated Germany and sought rocket-related employment as a writer and lecturer in Switzerland and Italy. In 1955, he joined von Braun's team of German rocket scientists at the U.S. Army's Redstone Arsenal. He worked there for several years before returning to Germany in 1958 and retiring.

Of the three founding fathers of astronautics, only Oberth lived to see some of his pioneering visions come true. These visions included: the dawn of the space age (1957), human space flight (1961), the first human landing on the Moon (1969), the first space station (1971), and the first flight of a reusable launch vehicle–NASA's space shuttle (1982). He died in Nuremberg, Germany, on December 29, 1989.

Oberth studied the theoretical problems of rocketry and outlined the technology needed for people to live and work in space. The last paragraph of his 1954 book *Man into Space*, addresses the important question: Why space travel? His eloquently philosophical response is: "This is the goal: To make available for life every place where life is possible. To make inhabitable all worlds as yet uninhabitable, and all life purposeful."

Under the technical leadership of von Braun, this program produced a number of experimental rocket vehicle designs, the most famous of which was the large, liquid-fueled A-4 rocket. The German military gave this rocket its more sinister and well-known name, V-2, for "vengeance weapon number two." It was the largest rocket vehicle at the time, at about 45.9 feet (14 m) long and 5.6 feet (1.7 m) in diameter and developing some 56,000 pounds-force (249,000 N) of thrust. The V-2 rocket had a liquid-propellant engine that burned alcohol and liquid oxygen.

At the end of World War II, the majority of the German rocket development team from the Peenemünde rocket test site, including von Braun,

Hermann J. Oberth was one of the cofounders of astronautics. Throughout his life, he vigorously promoted the concept of space travel. Unlike Robert Goddard and Konstantin Tsiolkovsky (the other founding fathers of astronautics), Oberth lived to see the arrival of the space age and human spaceflight, including the Apollo Project lunar landings. *(NASA)*

A reassembled German V-2 rocket lifting off its launch pad at the U.S. Army's White Sands Missile Range in southern New Mexico, circa 1947 *(U.S. Army/White Sands Missile Range)*

surrendered to the U.S. Army. This team of German rocket scientists and captured V-2 rocket components were then sent to White Sands Missile Range (WSMR), New Mexico, to initiate an American military rocket program. The first reassembled V-2 rocket was launched by a combined American-German team on April 16, 1946. A V-2, assembled and launched on this range, was America's first rocket to carry a heavy payload to high altitude. Another V-2 rocket set the first high-altitude and velocity record for a single-stage missile, and still another V-2 rocket was the first large missile to be controlled in flight.

✧ Cold–War Missile Race and the Arrival of the Space Age

Stimulated by the cold war, the American and German scientists at White Sands pressed on with the development of a variety of new military missile and rocket systems, including the Corporal, Redstone, Jupiter, and Nike Ajax.

The need for more room to fire rockets of longer range became evident in the late 1940s. In 1949 the Joint Long Range Proving Ground was established at a remote, deserted location on Florida's eastern coast known as Cape Canaveral. On July 24, 1950, a two-stage Bumper rocket became the first rocket vehicle to be launched from this now-famous location. The Bumper-8 rocket vehicle consisted of a V-2 first stage and a WAC-Corporal rocket second stage.

Within the former Soviet Union rocket engineers were also busily testing captured German V-2 rockets and beginning to make modifications and improvements, which led to new families of powerful liquid-propellant rockets. Rising superpower tensions between the Soviet Union and the United States translated directly into a great missile race and the dawn of the space age. The cold war era triggered tremendous developments in rocketry during the second half of the 20th century. On October 4, 1957, the Russian rocket engineer Sergei Korolev received permission to use a powerful military rocket to launch *Sputnik 1*—the first artificial satellite to orbit Earth.

The United States quickly responded on January 31, 1958, with the launching of *Explorer 1,* the first American satellite. The U.S. Army Ballistic Missile Agency (including von Braun's team of German rocket scientists, then located at the Redstone Arsenal in Huntsville, Alabama) modified a Redstone-derived booster into a four-stage launch vehicle configuration called the Juno I. This Juno I launch vehicle was the satellite-launching version of the army's Jupiter C rocket. The Jet Propulsion Laboratory (JPL) was responsible for the fourth stage, which included America's first

The first rocket launch from Cape Canaveral took place on July 24, 1950. The Bumper 8 rocket was a modified German V-2 rocket with a WAC Corporal second stage. *(United States Air Force)*

Late in the evening on January 31, 1958 (local time), this four-stage configuration of the Jupiter-C rocket blasted off from Cape Canaveral Air Force Station and successfully placed the *Explorer 1* spacecraft (the first United States satellite) into orbit around Earth. *(NASA)*

satellite. These historic launches marked the beginning of the age of space and the use of powerful rockets to send people and payloads beyond the cradle of planet Earth.

✧ To the Moon, Mars, and Beyond

During his brief term in office (January 20, 1961 to November 22, 1963), President John F. Kennedy continuously had to deal with conflicts involving the former Soviet Union, led by an aggressive premier, Nikita Khrushchev. The young president's challenges included serious confrontations over Cuba and Berlin, as well as a growing perception in the world community that the United States had lost its technical superiority to the Soviet Union. The Soviet premier constantly flaunted his nation's space technology accomplishments as evidence of the superiority of Soviet communism over Western capitalism. As had his predecessor President Dwight David Eisenhower (1890–1969), Kennedy worked hard to maintain a balance between American and Soviet spheres of influence in global politics. While not a space technology enthusiast per se, Kennedy recognized that its civilian space technology achievements were giving the former Soviet Union much greater influence in global politics. Driven by political circumstances early in his presidency, Kennedy took steps to respond to this challenge.

Space technology was the new, highly visible arena for Soviet-American competition. On April 12, 1961, the Soviet Union launched the first human (cosmonaut Yuri Gagarin) into orbit around Earth. The American response was a modest suborbital Mercury Project flight by astronaut Alan Shepard (on May 6, 1961). In May 1961, Kennedy boldly selected the Moon landing project. He did so not to promote space science or to satisfy a personal, long-term vision of space exploration, but because this mission was a truly daring project that would symbolize American strength and technical superiority in head-to-head cold-war competition with the Soviet Union.

Kennedy's decision gave NASA the mandate to expand and accelerate its Mercury Project activities and configure itself to accomplish the "impossible"

On July 16, 1969, American astronauts Neil Armstrong, Edwin (Buzz) Aldrin, and Michael Collins lifted off from Kennedy Space Center, Florida, onboard a gigantic Saturn V rocket on their way to the Moon. Four days later, two of these *Apollo 11* mission astronauts—Armstrong and Aldrin—became the first human beings to walk on another world. *(NASA)*

A Delta II expendable launch vehicle lifts off from Cape Canaveral Air Force Station, Florida, on June 10, 2003. This successful launch sent NASA's Mars Exploration Rover *Spirit* to the surface of the Red Planet—where (starting in January 2004) the robot rover began searching for scientific evidence whether past environments were once wet enough on Mars to be hospitable for life. *(NASA)*

through the Apollo Project. On July 16, 1969, a mammoth Saturn V rocket rumbled off its pad at Complex 39 of the Kennedy Space Center, Florida. It carried three Apollo astronauts (Neil Armstrong, Edwin "Buzz" Aldrin, and Michael Collins) on the most famous journey in history. Four days later—on July 20, two of these *Apollo 11* mission astronauts, Armstrong and Aldrin, walked on the surface of the Moon and successfully fulfilled Kennedy's bold initiative. Sadly, the young President who launched the most daring space exploration project of the cold war did not personally witness its triumphant conclusion. An assassin's bullet had taken his life in Dallas, Texas on November 22, 1963. NASA's Kennedy Space Center—site of Launch Complex 39, from which humans left Earth to explore the Moon—bears his name. Through Kennedy's bold and decisive leadership, human beings traveled through interplanetary space and walked on another world for the first time in history. The Apollo lunar landings were made possible by von Braun's mighty Saturn V launch vehicle, which was a direct descendent of the German V-2 rocket of World War II.

Since the late 1950s, rockets have enabled many other spectacular space missions. The first flight of NASA's space shuttle on April 12, 1981, opened up the era of aerospace vehicles and interest in reusable space transportation systems. With the end of the cold war in 1989, the space race was replaced by cooperative programs that are truly international in their perspective. For example, during the construction and operation of the *International Space Station,* both American and Russian rockets have carried people and equipment. Today, modern rocket technology has matured a long way from the gunpowder rockets of ancient China and Robert Goddard's first liquid-propellant rocket. In fact, seemingly routinely (although rocket launches are anything but routine) expendable rockets send sophisticated robot explorers to Mars and other fascinating worlds in our solar system. In this century, more advanced engineering versions of the rockets advocated by such space travel visionaries as Tsiolkovsky, Oberth, and Goddard will continue to serve as the dream machines that will help fulfill our destiny among the stars.

Rocket Propulsion Fundamentals

This chapter describes the physical principles upon which all rockets, large and small, operate. The basic features of the two major types of chemical rockets—solid propellant and liquid propellant—are also discussed.

In general terms, a rocket is a completely self-contained projectile propelled by a reaction engine. The reaction engine is a device that develops thrust by its physical reaction to the ejection of a substance. Typically, the reaction engine ejects a stream of hot gases created by combusting a propellant within the engine. Since it carries all the propellants it needs to operate, a rocket vehicle is a special type of reaction engine that can function in the vacuum of outer space. This unique capability makes the rocket the technical key to space travel.

✧ The Physical Principles behind the Rocket Engine

There are several types of thrust-producing reaction engines found in aerospace applications. These include the turbojet engine, the ramjet, and the rocket.

The basic turbojet engine is a reaction propulsion device that obtains the oxygen it needs to internally combust a hydrocarbon fuel from Earth's atmosphere. The incoming air is compressed by a turbine and then mixed with a finely sprayed fuel in the turbojet's burner or combustion chamber. Combustion produces hot gases, which are then expelled through a tail pipe and nozzle assembly to produce the thrust that propels an aircraft or cruise missile through the atmosphere. The compressor-turbine has a common shaft. One set of turbine blades (at the inlet of the turbojet) compresses the incoming air, while another set of blades at the other end of the shaft extracts some of the kinetic energy from the gases being expelled. This spins the compressor-turbine assembly as the jet-powered aircraft or cruise missile flies through the atmosphere. The design feature also gives rise to the term *turbojet*.

Like the rocket and the ramjet, the turbojet operates in accordance with Sir Isaac Newton's third law of motion. This important physical law states. "For every action there is an equal and opposite reaction." The major difference between the rocket and the turbojet is that the rocket is a self-contained unit that carries its own supply of both fuel and oxidizer (oxygen liberating material). The turbojet carries a supply of fuel, but obtains the oxygen necessary for combustion from the surrounding atmosphere. Because of this important difference, the rocket is not limited to operating just within the lower portions of Earth's atmosphere and can operate and produce thrust anywhere, including outer space.

The ramjet is another type of reaction engine that achieves jet propulsion in Earth's atmosphere. In principle, the ramjet is less complicated than a turbojet engine because it does not need the compressor-turbine assembly used in turbojet engines. Instead, the ramjet depends for its operation on the air compression as accomplished by the rapid forward motion of the vehicle. The ramjet has a specially shaped tube or duct open at both ends into which fuel is fed at a controlled rate. The air needed for combustion is shoved, or "rammed," into the duct and compressed by the forward motion of the vehicle/engine assembly. This rammed air passes through a diffuser and is mixed with fuel and burned. The combustion products are then expanded in and expelled through a nozzle. A ramjet cannot operate under static conditions. The duct geometry depends on whether the ramjet operates at subsonic or supersonic speeds. Often, a rocket is used to achieve the initial forward motion of the ramjet; such devices are called rocket-ramjets.

While simple in concept, maintaining combustion within high-speed aerodynamic flow conditions that often include shock wave formation is not a trivial design task. Aerospace engineers are now examining the use of hypersonic ramjet vehicles as part of future space transportation systems. Unlike the rocket, the ramjet needs to extract the oxygen for propellant combustion from Earth's atmosphere. Future hypersonic ramjets could be capable of operating at very high speeds in the upper regions of Earth's atmosphere. If such reaction engines are constructed in the next few decades, they could serve as the first propulsive stage of a future reusable aerospace launch vehicle system.

The rocket is the third type of propulsive reaction engine and is the central theme of this book. Since it carries all of its propellant and operates independently of its environment, the rocket works in outer space and represents the key to space travel. Aerospace engineers often find it convenient to classify rockets by the energy source they use to accelerate the ejected matter, which creates the vehicle's thrust. For example, engineers talk about chemical rockets, nuclear rockets, and electric rockets. This chapter deals with the general characteristics and operational principles of chemical rockets. Nuclear rockets are described in chapter 9 and electric rockets in chapter 10. Chemical rockets, in turn, often are divided into two general subclasses: solid-propellant rockets and liquid-propellant rockets.

✧ Rockets and Newton's Laws

The rocket, as well as other reaction engines, operates in accordance with Newton's third law of motion—that is, the action-reaction principle. Although the physical principle was observed and used since ancient times, the scientific basis for this important physical law was not formally identified until the brilliant British physicist and mathematician Sir Isaac Newton published his great work, *Principia Mathematica,* in 1687.

One of the first devices to successfully demonstrate the principles essential to the performance of a rocket was a toy wooden bird. The writings of the Roman author and grammarian Aulus Gellius (ca.130–180 C.E.) mention an ancient Greek named Archytas, who lived about 400 B.C.E. in the city of Tarentum in southern Italy. Archytas amused the children of Tarentum with a toy wooden pigeon that flew through the air suspended on a wire as it was propelled by escaping steam.

Similarly, about 60 C.E., the Greek engineer and mathematician Hero of Alexandria constructed his aeoliphile. This device was a spinning steam-powered toy that clearly showed the action-reaction principle. In Hero's aeoliphile, a fire below the hollow copper sphere turned water into steam, which then served as a propulsive gas. The steam escaped through two L-shaped tubes, mounted on opposite sides of the copper sphere. The escaping steam passed through these two crude nozzles and provided thrust to the sphere, causing it to rotate rapidly.

Since the 13th century, people used gunpowder rockets for amusement, as well as for military weapons. But it is only in the last three centuries that scientists have understood the physical basis for their operation. In a very real sense, rocketry, as a science, began in about 1687, when Newton published his famous book, *Principia Mathematica.* In this brilliant work, Newton stated three important scientific principles that describe the motion of almost any object, including rockets. Knowing these basic principles, called Newton's laws of motion, aerospace engineers can predictably design the powerful rockets that deliver spacecraft and human crews into space.

SIR ISAAC NEWTON
(1642–1727)

Sir Isaac Newton was the brilliant, though introverted, British astrophysicist and mathematician whose law of gravitation, three laws of motion, development of calculus, and design of a new type of reflecting telescope make him one of the greatest scientific minds in human history. Newton's great work, *Philosophiae Naturalis Principia Mathematica* (Mathematical principles of natural philosophy; also known as *Principia Mathematica*), appeared in 1687. This monumental book transformed the practice of physical science and completed the scientific revolution started by Nicolaus Copernicus (1473–1543), Johannes Kepler (1571–1630), and Galileo Galilei (1564–1642).

Newton was born prematurely in Woolsthorpe, Lincolnshire, on December 25, 1642. His father had died before Newton's birth and this event contributed to a very unhappy childhood. Throughout his life, Newton would not tolerate criticism, remained hopelessly absent-minded, and often tottered on the verge of emotional collapse. British historians claim that Newton laughed only once or twice in his entire life. Yet he is still considered by many science historians to be the greatest human intellect who ever lived.

He graduated without any particular honors or distinction from Cambridge in 1665, earning a bachelor's degree. Following graduation, Newton returned to his mother's farm in Woolsthorpe to avoid the plague that had broken out in London. This self-imposed exile laid the foundation for his brilliant contributions to science. By Newton's own account, one day on the farm he saw an apple fall to the ground and began to wonder if the same force that pulled on the apple also kept the Moon in its place. In Newton's time, heliocentric cosmology was becoming widely accepted (except where banned on political or religious grounds), but the mechanism for planetary motion around the Sun remained unexplained.

By 1667, the plague epidemic subsided and Newton returned to Cambridge as a minor fellow at Trinity College. The following year he received his master of arts degree and became a senior fellow. In about 1668, he constructed the first working reflecting telescope, a device that now carries his name. The Newtonian telescope uses a parabolic mirror to collect light. The primary mirror then reflects the collected light by means of an internal secondary mirror to an external focal point at the side of the telescope's tube. This new telescope design earned Newton a great deal of professional acclaim, including eventual membership in the Royal Society.

In 1669, Isaac Barrow, Newton's former mathematics professor, resigned his position so that the young Newton could succeed him as Lucasian Professor of Mathematics. This position provided Newton the time to collect his notes and properly publish his work—a task he was always tardy to perform.

Shortly after his election to the Royal Society (in 1671), Newton published his first paper in that society's transactions. While an undergraduate, Newton had used a prism to refract a beam of white light into its primary colors (red, orange, yellow, green, blue, and violet). Newton reported this important discovery to the Royal Society. But

Newton's first law of motion introduces the concept of inertia. It states that objects at rest will stay at rest and objects moving in a straight line will keep moving in a straight line, unless acted upon by an unbalanced external

Robert Hooke (1635–1703)—an influential member of the society—immediately attacked Newton's pioneering investigation into the nature of light.

This was the first in a lifelong series of bitter disputes between Hooke and Newton. Newton only skirmished lightly then quietly retreated. This was Newton's lifelong pattern of avoiding direct conflict. When he became famous later in his life, Newton would start a controversy, withdraw, and then secretly manipulate others who would then carry the brunt of the battle against Newton's adversary. For example, Newton's famous conflict with the German mathematician Gottfried Leibniz (1646–1716) over the invention of calculus followed precisely such a pattern. Through Newton's clever manipulation, the calculus controversy even took on nationalistic proportions, as carefully coached pro-Newton British mathematicians bitterly argued against Leibniz and his supporting group of German mathematicians.

In August 1684, Edmund Halley (1656–1742) made a historic trip to visit Newton at Woolsthorpe and convinced the reclusive genius to address the following puzzle about planetary motion. In part to fulfill his promise, Newton sent Halley his *De Motu Corporum in Gyrum* (On the motion of bodies in orbit). In this document, Newton demonstrated that the force of gravity between two bodies is directly proportional to the product of their masses and inversely proportional to the square of the distance between them. Physicists now call this important relationship Newton's universal law of gravitation. Halley was astounded and begged Newton to carefully document all of his work on gravitation and orbital mechanics. Through Halley's patient encouragement and financial support, Newton eventually published *Principia Mathemat-*

ica in 1687. In this book, Newton gave the world his famous three laws of motion and the universal law of gravitation. Newton's monumental work transformed physical science and completed the scientific revolution started by Copernicus, Kepler, and Galileo. Many consider the *Principia* as the greatest scientific accomplishment of the human mind.

After completing the *Principia*, Newton drifted away from physics and astronomy and eventually suffered a serious nervous disorder in 1693. Upon recovery, he left Cambridge in 1696 to assume a government post in London as warden (then later master) of the Royal Mint. During his years in London, Newton enjoyed power and worldly success. Robert Hooke, his lifelong scientific antagonist, died in 1703. The following year (1704), the Royal Society elected Newton its president. Unrivaled, he won annual reelection to this position until his death. Newton was so bitter about his quarrels with Hooke that he waited until 1704 to publish his other major work, *Opticks*. Queen Anne knighted him in 1705.

Although his most innovative years were now clearly far behind him, Newton still continued to exert great influence on the course of modern science. He used his position in the Royal Society to exercise autocratic control over the careers of many younger scientists. Though highly esteemed, Newton remained intolerant of controversy. So, as society president, he skillfully maneuvered younger scientists to fight his intellectual battles. In this almost tyrannical manner, he continued to rule the scientific landscape until his death in London on March 20, 1727. To honor his great achievements, scientists and engineers around the world now call the SI unit of force the newton (N).

force. Physicists also call this statement the conservation of linear momentum principle. In physics, the linear momentum of an object (p) is equal to the product of the object's mass (m) and its velocity (v)—that is, p = mv.

Newton's second law states that the rate of change of momentum of a body is proportional to the force acting upon the body and is in the direction of the applied external force. For an object of constant mass, this results in the familiar statement: Force (F) is equal to mass (m) times acceleration (a)—that is, F = ma. During their propellant-burning operations, chemical rockets continuously undergo a loss of mass, so rocket engineers must use a slightly more complicated mathematical form of Newton's second law, which takes this continuously decreasing mass into account. However, a detailed discussion of this more complex equation involves an understanding of calculus and lies beyond the scope of this book.

Newton's third law of motion is the action-reaction principle, which represents the physical basis of all rockets. It states that for every force acting upon a body, there is a corresponding force of the same magnitude that the body exerts in the opposite direction.

These deceptively simple statements form the basis of Newtonian mechanics, an extremely powerful and useful tool in science and engineering. In Newtonian mechanics, mass and energy are treated as separate, conservative mechanical properties. Early in the 20th century, another brilliant scientist, Albert Einstein (1879–1955), extended Newtonian mechanics into the realm of relativistic physics. Einstein did this by treating mass and energy as equivalent with his famous equation, $E = mc^2$. In chapter 12, this book examines Einstein's work in relativity within the context of very exotic advanced propulsion system concepts. Until then, Newtonian mechanics adequately explains how powerful (but nonrelativistic) rocket vehicles work. Because of their very special importance in rocketry, this chapter now discusses each of Newton's laws of motion in a little more detail.

NEWTON'S FIRST LAW OF MOTION

This law is simply a statement of an easily observable physical fact. But to really know what it implies, the scientific meaning of the terms *rest, motion,* and *external unbalanced force* must be appreciated. Think of rest and motion as opposite of each other. Rest is the state of an object when it is not changing position in relation to its surroundings. For example, if a person is sitting still in a chair, he (or she) is at rest with respect to the chair.

Now, assume the chair in which the person is sitting motionless is actually one of many seats on a high-flying commercial jet plane. Relative to Earth's surface, that person and the chair are certainly moving. However, he is still at rest with respect to the chair, since he is not moving in relation to his immediate surroundings. If scientists defined rest "as the total absence of motion," such a definition would really be meaningless, because this condition does not actually exist in nature. Even if the person was sitting comfortably in a chair at home (and not on a jet plane), he (or she) would still be moving, because the chair is located on the surface of a spinning planet that is orbiting a star. That star (the Sun) is also moving through space, rotating around a galaxy that is itself moving through the universe. So while a person is sitting "still" at home, he is actually traveling through space at a speed of many kilometers per second.

Of course, humans do not feel this motion, and for practical problem-solving here on Earth, scientists and engineers often "neglect" the planet's motion through space. In aerospace engineering, as well as other technical fields, it is important to always understand the assumptions and constraints that are inherent in a particular scientific model of the physical world. Planetary motions do become important in the field of astronautics, when, for example, rocket scientists plan to send a spacecraft from Earth to Mars.

Motion is also a relative term. All matter in the universe is moving all the time. However, in the context of Newton's first law, the term *motion* specifically means that an object is changing its position in relation to its immediate surroundings. A ball is at rest if it lays motionless on the ground. But when it is rolling on the ground, scientists say the ball is in motion. The rolling ball keeps changing its position in relation to its immediate surroundings. Similarly, a rocket vehicle blasting off its launch pad changes from a state of rest to a state of motion.

The third term necessary to understand Newton's first law is that of an external, unbalanced force. External means that the force comes from outside the object. If a person holds a ball in the palm of one of their hands and keeps that palm open upward and motionless, the ball remains at rest with respect to both the hand and the immediate surroundings. While the person holds the ball motionless in the air, the force of Earth's gravity keeps trying to pull the ball downward. Why doesn't it move? The ball does not move because the person's hand provides a lift force that pushes just the right amount against the ball to hold it up and motionless. In this case, the external forces acting on the ball are balanced. The downward force of gravity and the lift force provided by

In about 60 C.E., the Greek engineer and mathematician Hero of Alexandria created the *aeoliphile* (shown here)—a toylike device that demonstrates the action-reaction principle, which is the basis of operation of all rocket engines.

hollow sphere

steam

© Infobase Publishing

the person's hand keep the ball suspended in the air. But if the person quickly tilts that hand and lets the ball go, the external forces on the ball become unbalanced. Now, the unopposed force of gravity makes the ball drop to the ground. In this case, the ball changes from a state of rest to a state of motion. In the late 16th and early 17th centuries, the brilliant Italian scientist Galileo Galilei performed many pioneering studies concerning the motion of objects under the influence of Earth's gravity. Newton built upon Galileo's work involving the physics of free fall and the ballistic motion of projectiles to construct his three laws of motion.

During the flight of a rocket, forces change between balanced and unbalanced all the time. In the countdown, for example, a sounding rocket rests motionless on its launch pad and all the external forces on the rocket are balanced. The force of gravity pulls it downward, while the surface of the launch pad pushes it upward. When the countdown reaches zero, the launch director sends a special command signal that ignites the rocket's engine. Now, the thrust from the engine creates an unbalanced force and the rocket travels upward. Later, when its propellant supply runs out, the sounding rocket slows down, stops at the highest point in its flight, and then starts falling back to Earth.

NEWTON'S SECOND LAW OF MOTION

This important physical law is essentially a statement of the equation that force (F) equals the mass (m) times the acceleration (a), that is, $F = m \cdot a$. To explore the great significance of this physical principle, scientists often use the old-fashioned cannon shown in the figure. When this cannon is fired, the explosive charge releases gases that propel the cannon ball out the open end of the barrel. Depending on the mass of the cannon ball and the energy released in the explosion, the cannon ball might travel a mile (km) or so to its target. At the same time, the cannon itself recoils (jumps backward) about a foot (m) or two. (The recoil action happens because of Newton's third law.) The explosive force acting on the cannon and the cannon ball is the same. Newton's second law describes what happens to each object after the explosion. Scientists and engineers like to perform "second law analyses" to understand how objects interact with applied forces. So they first write the second law for each object. The first equation, $F_{explosion} = M_{(cannon)} A_{(cannon)}$, refers to the cannon, while the second equation, $F_{explosion} = M_{(ball)} A_{(ball)}$, pertains to the cannon ball. Now, in the first equation, the mass is that of cannon ($M_{(cannon)}$) and the acceleration ($A_{(cannon)}$) involves the recoil movement of the cannon after the explosion and departure of the cannon ball. In the second equation, the mass is that of the cannon ball ($M_{(ball)}$) and the acceleration ($A_{(ball)}$) relates to its high-

This old-fashioned cannon helps explain Newton's second law of motion.

© Infobase Publishing

speed departure. Assuming the explosive force is the same for both equations, physicists then combine and rewrite these equations as follows:

$$M_{(cannon)} \, A_{(cannon)} = M_{(ball)} \, A_{(ball)}$$

In order to keep both sides of this new equation balanced, the accelerations experienced by the cannon and the cannon ball must vary with their respective masses. As a result, the cannon with its very large mass, experiences a small, modest acceleration, while the cannon ball with its much smaller mass experiences a very large acceleration, during the explosive event.

To help themselves understand how the second law describes the performance of a rocket engine, physicists often imagine placing this same cannon somewhere in interplanetary space. The cannon now represents the mass structure of a very unusual "rocket ship" and the cannon ball represents the total mass of the combustion gases ejected to space through the nozzle of this unusual "rocket ship." Scientists next assume that the explosive charge within the cannon represents the high-pressure thermodynamic conditions created by the combustion of the chemical propellants inside a chemical rocket's engine. In this simple model of rocket performance, the "exhaust gases" (here represented by the cannon ball) leave the rocket at very high acceleration and the "rocket ship" (here the cannon) responds by accelerating or recoiling in the opposite direction.

A few important features of rocket-engine design should become apparent from this simple scenario. The more cannon balls that are fired, the more total acceleration given to the recoiling cannon (that is, the rocket ship). The lower the total mass of the cannon (rocket ship), the greater its final acceleration. Finally, the greater the acceleration of each cannon ball fired, the greater the total recoil acceleration of the cannon (rocket ship). These are precisely the reasons why rocket engineers try to make the mass of the rocket vehicle (including the payload) as low as possible, react as much high-energy propellant as possible, and expand the combustion gases through an appropriately designed nozzle at the highest-possible exit velocity.

Of course, this simple scenario represents only an instantaneous picture (snapshot) of how a rocket engine really works. It does not show other interesting things that happen within real rockets. Perhaps the most important difference is the fact that when the cannon is fired and the ball flies out, the entire event lasts only a brief moment and produces an instantaneous, impulsive (reaction) thrust. For chemical rockets, the generation of thrust is actually a reasonably continuous process that lasts as long as the engine has propellant to burn (typically many seconds to several minutes). A second important difference is that the mass of the rocket keeps changing during the powered (thrusting) portion of its flight. The rocket vehicle's mass is the sum of all its parts, including: engine, propellant tanks, payload, control system, and propellants. By far the largest part of the chemical rocket's mass is its propellants. (Typically, 85 to 92 percent of the total takeoff mass of a modern rocket is propellant.) Finally, the amount of propellant constantly changes (decreases) as the engine fires. That is why a rocket vehicle starts off climbing slowly and then goes faster and faster as it ascends into space.

It is helpful to restate Newton's second law within the context of rocket science as follows: The greater the mass of rocket fuel burned, and the faster the combustion gases can escape through the nozzle, the greater the thrust of the rocket. Rocket engineers sometimes use the specific impulse (I_{sp}) as a figure of merit when they want to compare the performance of different rocket engines.

SPECIFIC IMPULSE (SYMBOL: I_{SP})

The specific index is an important concept that rocket scientists use as a performance index for rocket propellants. It is defined as the thrust (or thrust force) produced by propellant combustion divided by the propellant mass flow rate. Expressed as an equation, the specific impulse is:

$$I_{sp} = \text{thrust / mass flow rate}$$

It is also helpful to understand the units associated with specific impulse. In the SI unit system, thrust is expressed in newtons and mass flow rate in kilograms per second. Since one newton equals one kilogram-meter per second-squared (i.e., $1\ N = 1\ kg\text{-}m/s^2$), the specific impulse in SI units becomes:

$$I_{sp} = \text{newtons / (kg/sec)} = \text{meter/second (m/s)}$$
[in SI units]

Sometimes aerospace engineers in the United States use the traditional (or American) engineering system of units. In this unit system, thrust is expressed in pounds-force (lbf), while mass flow rate of propellant in pounds-mass per second (lbm/s). Since by definition within this unit system, one pound-force is equal to one pound-mass at sea level on the surface of Earth, aerospace engineers often use the following simplification—which is, strictly speaking, valid only at sea level on Earth:

$$I_{sp} = \text{lbf/(lbm/s)} = \text{seconds (s)} \text{ [in American engineering units]}$$

This simplification often causes a great deal of confusion when rocket engineers try to describe the performance of a particular rocket engine using both systems of units. For example, a modern liquid propellant chemical rocket that burns hydrogen (as the fuel) and oxygen (as the oxidizer) will have a theoretical specific impulse between 2,940 to 3,725 m/s in the SI unit system and between 300 and 380 s in the American engineering unit system.

NEWTON'S THIRD LAW OF MOTION

This physical law states that for every action there is an equal and opposite reaction. The action-reaction principle is the basis of operation of all rockets. The rocket engine expels mass at high velocity and the reaction thrust drives the rocket vehicle in the opposite direction.

The thrust equation is the fundamental equation for rocket engine performance. For reaction engines, which generate thrust by expelling a stream of internally carried mass, rocket scientists write this important equation as:

$$T = \dot{m}\, V_e + (p_e - p_a)\, A_e$$

where T is the thrust or reaction force (expressed in pounds-force [N]), $\dot{m}$ is the mass flow rate of ejected materials (pounds-mass per second [kg/s]), V_e is the exhaust velocity of the ejected mass (feet per second (m/s), p_e is the exhaust pressure at the nozzle exit (pounds-force per foot-squared [N/m^2]), p_a is the ambient pressure (pounds-force per foot-squared [N/m^2]), and A_e is the nozzle exit area (square feet [m^2]).

Many people make the common mistake of believing that a rocket is propelled through the air by its exhaust gases pushing against the outside air. Nothing could be farther from the truth. As initially demonstrated by Robert H. Goddard, rockets work better in outer space, where the ambient pressure (p_a) is zero.

While this brief look at Newton's laws of motion may seem a bit overwhelming, these physical laws provide a basic understanding of how a rocket really works. An unbalanced force must be applied to a rocket, if it is to rise from the launch pad and climb into space. The same is true, if a space vehicle changes its speed or direction, while traveling through space (first law). The rate at which a rocket consumes its propellant mass and the exhaust speed of the ejected materials determine the amount of thrust (force) produced by the rocket engine (second law). The reaction (or forward) motion of a rocket vehicle is equal to and opposite of the action (or thrust) from the engine (third law).

✦ Basic Rocket Science

A rocket vehicle is a completely self-contained device propelled by a reaction engine. In its simplest form, the rocket can be considered as just a chamber enclosing a propellant gas under pressure. A small opening at one end of the chamber, called a nozzle, allows this pressurized gas to escape. As the gas rushes out through the nozzle, it produces a reaction thrust that propels the rocket in the opposite direction.

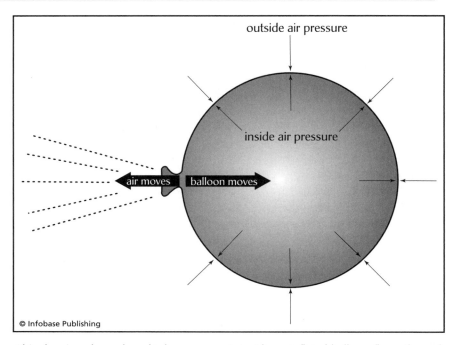

outside air pressure

inside air pressure

air moves | balloon moves

© Infobase Publishing

This drawing shows how high-pressure air inside an inflated balloon flows through an opening (nozzle) into the surrounding lower pressure air, generating a reaction thrust that moves the balloon to the right.

An inflated balloon provides a simple way to understand this concept a little better. Within its elastic limits, a balloon's stretched rubber wall confines the high-pressure air that keeps the balloon inflated. When completely sealed, the inflated balloon's wall pushes in and the confined air pushes back so that the inward force (stretched balloon wall) and outward force (internal air pressure) balance. Physicists call this balanced condition, a state of mechanical equilibrium. However, if the pressurized air is suddenly released through a narrow opening (nozzle) in the balloon, the air rushes away from the high-pressure region inside and escapes to the lower-pressure environment outside. As the air flows through the nozzle, a reaction force occurs that propels the balloon in the opposite direction. Often when a person tries to blow up a balloon for a party, it slips away at the last minute and then scoots around the room. Such an erratic flying balloon obeys the same basic laws as the powerful rockets that send spacecraft into orbit.

Of course, there is a major difference in the way rockets and balloons acquire the thrust-producing pressurized gas. A chemical rocket generates this high-pressure gas by burning propellants inside its combustion chamber. In the case of a balloon, a person provides the pressurized gas by pumping (or blowing) air into an elastic (usually latex) enclosure that expands as the air pressure increases up to some safe limit.

Aerospace engineers must avoid designing rockets that have a lot of unnecessary mass. In fact, they do everything they can to slim down a rocket vehicle to the bare essentials. This often involves using specially created low-mass materials, clever structural designs that are low mass but do not compromise the vehicle's integrity under operational loads, and staging—the process of discarding useless mass during the flight—as clever ways of improving a rocket's mass fraction (MF). The figure below provides an exploded view of the Delta II expendable launch vehicle that sent NASA's *Stardust* spacecraft on its successful rendezvous mission with comet Wild 2. Take special note of the numerous rocket stages that were jettisoned as the vehicle ascended into space.

Rocket scientists define MF as the mass of the propellants a rocket carries divided by its total mass (including propellant load). This definition

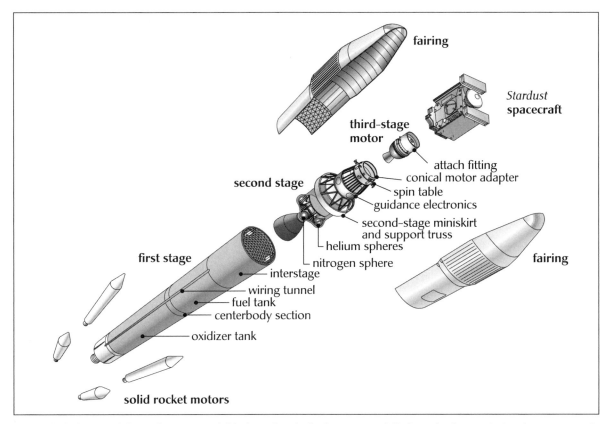

An exploded view of the Delta II expendable launch vehicle that successfully launched NASA's *Stardust* spacecraft from Cape Canaveral, Florida, in early February 1999

means that if the entire rocket is nothing more than a giant lump of propellant, its MF has a value of unity. Of course, that would be a bit senseless, because that type of rocket would not carry any payload. Whether aerospace engineers are building a solid-fueled or liquid-fueled rocket, there are some physical limits on the minimum mass of the structural components and the control hardware that we must use to contain and then burn a given mass of propellant. Engineers consider an "ideal chemical rocket" to be one for which the total mass of the vehicle is distributed roughly as follows: propellants, 91 percent of total mass; structure and control hardware (including engines, tanks, casings, fins, pumps, etc), 3 percent of the total initial mass; and payload, 6 percent of the total initial mass. The larger the MF value, the smaller the amount of payload a rocket vehicle can carry. In addition, the smaller the MF value, the shorter a rocket's range becomes, because of propellant supply limitations.

Some rockets are used in military missions to carry warheads (explosive payloads) against enemy targets, near and far away. An intercontinental ballistic missile (ICBM) has a range of 3,400 miles (5,500 km) or more. Aerospace engineers consider an ICBM with an MF number of 0.91 to represent a good balance between payload-carrying capability and range. As a point of reference, NASA's space shuttle has a mass fraction of 0.82, although this number varies somewhat from mission to mission and orbiter vehicle to orbiter vehicle.

Large rockets, the kind needed to carry spacecraft into orbit, have serious mass fraction problems. To achieve orbital or escape velocities, these large rockets must consume a great deal of propellant. The hardware to carry and burn all this propellant becomes excessively massive as the rocket vehicle becomes larger. Why carry all the extra structural mass, when a great deal of the propellant is gone? The answer is simple! Do not! Engineers actually design modern rockets to discard mission-useless mass as the vehicle climbs in altitude. Aerospace engineers call the process staging, and modern "step rockets" have opened outer space to exploration by both robot spacecraft and human crews.

The first stage rocket of a step rocket is the largest, because it must propel itself as well as all the companion upper stages to some altitude and velocity. Upon depleting its propellant supply, the first stage then separates and falls away from the rest of the flight vehicle. Next, the second-stage engine fires. The process continues with excess structural mass being discarded at every step in the sequence. Eventually, the payload reaches orbital (or even escape) velocity in an efficient (favorable mass fraction) manner.

Most modern rockets operate with either solid or liquid chemical propellants. The term *propellant* does not simply mean fuel; it refers to both fuel and oxidizer. The fuel is the chemical propellant the rocket engine burns, but an oxidizer is also needed to supply the oxygen necessary for combustion.

✧ Solid–Propellant Chemical Rockets

A solid-propellant chemical rocket is the simplest kind of rocket. This type of rocket traces its technical heritage all the way back to the gunpowder-fueled fire arrows of ancient China. Those in use today generally consist of a solid propellant (i.e., fuel and oxidizer compound) with the following associated hardware: case, nozzle, insulation, igniter, and stabilizers.

Solid propellants, commonly referred to as the "grain," are basically a chemical mixture or compound containing a fuel and oxidizer that burn (combust) to produce very hot gases at high pressure. The important feature here is that these propellants are self-contained and can burn without the introduction of outside oxygen sources (such as air from Earth's atmosphere). Consequently, the solid-propellant rocket, and its technical sibling, the liquid-propellant rocket (discussed later), can operate in outer space.

Solid propellants often are divided into three basic classes: monopropellants, double-base, and composites. Monopropellants are energetic compounds such as nitroglycerin or nitrocellulose. Both of these compounds contain fuel (carbon and hydrogen) and oxidizer (oxygen). Monopropellants are rarely used in modern rockets. Double-base propellants are mixtures of monopropellants, such as nitroglycerin and nitrocellulose. The nitrocellulose adds physical strength to the grain, while the nitroglycerin is a high-performance, fast-burning propellant. Usually, double-base propellants are mixed together with additives that improve the burning characteristics of the grain. The mixture becomes a puttylike material that is loaded into the rocket case.

Composite solid propellants are formed from mixtures of two or more unlike compounds that by themselves do not make good propellants. Usually one compound serves as the fuel and the other as the oxidizer. For example, the propellants used in the solid rocket boosters (SRBs) of the space shuttle fall into this category. The propellant type used in this case is known as PBAN, which stands for polybutadiene acrylic acid acrylonitrile terpolymer. This somewhat exotic-sounding chemical compound is used as a binder for ammonium perchlorate (oxidizer), powdered aluminum (fuel), and iron oxide (an additive). The cured propellant looks and feels like a hard rubber eraser.

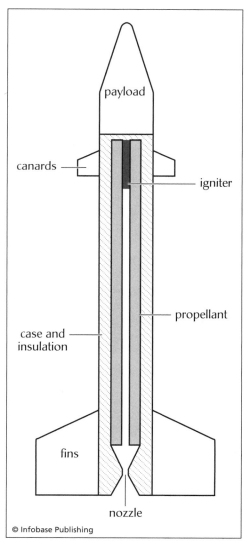

© Infobase Publishing

The basic components of a solid-propellant rocket

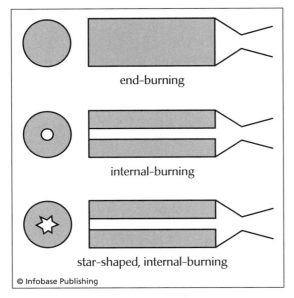

Typical solid-rocket motor grain designs

The thrust produced by the propellants is determined by the combustive nature of the chemicals used and by the shape of their exposed burning surfaces. A solid propellant will burn at any point that is exposed to heat or hot gases of the right temperature. Grains usually are designed to be either end-burning or internal-burning.

End-burning grains burn the slowest of any grain design. The propellant is ignited close to the nozzle, and the burning proceeds the length of the propellant load. The area of the burning surface is always at a minimum. While the thrust produced by end-burning is lower than for other grain designs, the thrust is sustained over longer periods.

Much more massive thrusts are produced by internal burning. In this design, the grain is perforated by one or more hollow cores, which extend the length of the case. With an internal-burning grain, the burning surface of the exposed cores is

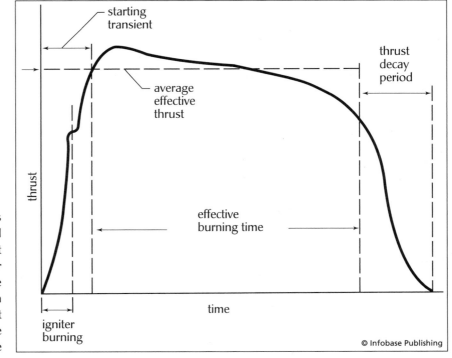

The thrust versus time plot for a typical solid-propellant rocket motor designed to produce a relatively uniform average thrust over the effective burning time

much larger than the surface exposed in an end-burning grain. The entire surfaces of the cores are ignited at the same time, and burning proceeds from the inside out. To increase the surface available for burning, a core may be shaped into a cruciform or star design.

By varying the geometry of the core design, the thrust produced by a large internal-burning grain also can be customized as a function of time to accommodate specific mission needs. The figure at the bottom of page 40 illustrates a thrust-versus-time plot for a solid-propellant motor with a grain design that provides a relatively uniform amount of thrust over the effective burning time of the rocket engine. The massive SRBs used by the space shuttle feature a single core that has an 11-point star design in the forward section. At 65 seconds into the launch, the star points are burned away and thrust temporarily diminishes. This coincides with the passage of the space shuttle vehicle through the sound barrier. Buffeting occurs during this passage and the reduced SRB thrust helps alleviate strain on the vehicle.

The rocket case is the pressure and load-carrying structure that encloses the solid propellant. Cases are usually cylindrical, but some are spherical in shape. The case is an inert part of the rocket, and its mass is an important factor in determining how much payload the rocket can carry and how far it can travel. Efficient, high-performance rockets require that the casing be constructed of the lightest materials possible. Alloys of steel and titanium often are used for solid rocket casings. Upper-stage vehicles may use thin metal shells that are wound with fiberglass for extra strength.

Unless protected by insulation, the solid rocket motor case will lose strength rapidly and burst or burn through. Therefore, to protect the casing, insulation is bonded to the inside wall of the case before the propellant is loaded. The thickness of this insulation is determined by its thermal properties and how long the casing will be exposed to the high-pressure, very hot combustion gases. A frequently used insulation for solid rockets is an asbestos-filled rubber compound that is thermally bonded to the casing wall.

During the combustion process, the resulting high temperature and high-pressure gases exit the rocket through a narrow opening, called the nozzle, which is located at the lower end of the motor. The most efficient nozzles are convergent-divergent designs. During operation, the exhaust gas velocity in the convergent portion of the nozzle is subsonic. The gas velocity increases to sonic speed at the throat and then to supersonic speeds, as it flows through and exits the divergent portion of the nozzle. The narrowest part of the nozzle is the throat. Escaping gases flow through this constricted region with relatively high velocity and temperature.

Excessive heat transfer to the nozzle wall at the throat is a great problem. Often thermal protection for the throat consists of a liner that either withstands these high temperatures (for a brief period of operation) or else ablates (i.e., intentionally erodes, carrying heat away). Large solid rocket motors, such as the space shuttle's solid rocket booster, generally rely on ablative materials to protect the nozzle's throat. Smaller solid rocket motors, such as might be used in an air-to-air combat missile or in a short-range surface-to-surface military missile, often use high-temperature-resistant materials to protect the nozzle. These temperature-resistant materials (possibly augmented by a thin layer of heat-resistant liner) can protect the nozzle's throat sufficiently, since the rocket's burn period is quite short (typically a few seconds).

To ignite the propellants of a solid rocket, the grain surface must be saturated with hot gases. The igniter is usually a rocket motor itself, but much smaller in size. The igniter can be placed inside the upper end of the hollow core, at the lower end of the core, or even completely outside the solid rocket motor. In the latter case, the exhaust of the igniter is directed into the nozzle of the larger solid rocket motor. An electrical circuit with a hot-wire resistor or an exploding bridgewire starts the igniter. The initial part of the ignition sequence begins with fast-burning pellets that, in turn, fire up the main igniter propellants.

Active directional control of solid-propellant rockets in flight generally is accomplished by one of two basic approaches. First, fins and, possibly, canards (small fins on the front end of the casing) can be mounted on the rocket's exterior. During flight in the atmosphere, these structures tilt to steer the rocket in much the same way that a rudder operates on a boat. Canards have an opposite effect on the directional changes of the rocket from that produced by the (tail) fins. Second, directional control also may be accomplished by using a gimbaled nozzle. Slight changes in the direction of the exhaust gases are accomplished by moving the nozzle from side to side. Large solid rocket motors, such as the space shuttle's solid rocket boosters, use the gimbaled-nozzle approach to help steer the vehicle.

© Infobase Publishing

Rocket scientists can use the controlled swiveling of a gimbaled nozzle to help steer a rocket vehicle.

Compared to liquid-propellant rocket systems, solid-propellant rockets offer the advantage of simplicity and reliability. With the exception of stability controls, solid-propellant rockets have no moving parts. When loaded with propellant and erected on a launch pad (or placed in an underground strategic missile silo or inside a launch tube in a ballistic missile submarine), solid rockets stand ready for firing at a moment's notice. In contrast, liquid-propellant rocket systems require extensive pre-launch preparations.

Solid rockets generally have an additional advantage in that a greater portion of their total mass can consist of propellants. Liquid-propellant rocket systems require fluid feed lines, pumps, and tanks, all of which add additional (inert) mass to the vehicle.

The principal disadvantage of solid-propellant rockets involves the burning characteristics of the propellants themselves. Solid propellants are generally less energetic (i.e., they deliver less thrust per unit mass consumed) than the best liquid propellants. Also, once ignited, solid-propellant motors burn rapidly and are extremely difficult to throttle or extinguish. In contrast, liquid-propellant rocket engines can be started and stopped at will.

Today, solid-propellant rockets are used for strategic nuclear missiles (e.g., the U.S. Air Force's Minuteman), for tactical military missiles (e.g., Stinger), for small expendable launch vehicles (e.g., Scout), and as strap-on solid boosters for a variety of liquid-propellant launch vehicles, including the reusable space shuttle and the expendable Titan-IV. Solid-rocket motors also are used in small sounding rockets and in many types of upper-stage vehicles, such as the Inertial Upper Stage (IUS) system.

✧ Liquid–Propellant Chemical Rockets

The American physicist Robert H. Goddard invented the liquid-propellant rocket engine in 1926. So important were his contributions to the field of rocketry that it is appropriate to state: "Every liquid propellant rocket is essentially a Goddard rocket."

The figure at the top of page 44 describes the major components of a typical liquid-propellant chemical rocket. Here, the propellants (liquid hydrogen for the fuel and liquid oxygen for the oxidizer) are pumped to the combustion chamber, where they begin to react. The liquid fuel often is passed through the tubular walls of the combustion chamber and nozzle to help cool them and prevent high-temperature degradation of their surfaces.

Liquid-propellant rockets have three principal components in their propulsion system: propellant tanks, the rocket engine's combustion chamber and nozzle assembly, and turbopumps. The propellant tanks

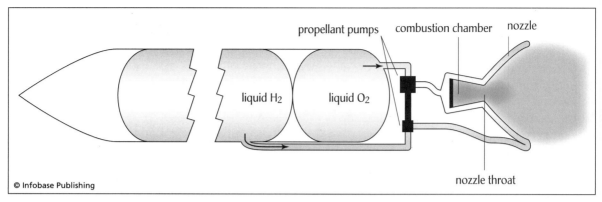

Basic hardware associated with a bipropellant (here LH_2 and LO_2) liquid rocket

are load-bearing structures that contain the liquid propellants. There is a separate tank for the fuel and for the oxidizer. The combustion chamber is the region into which the liquid propellants are pumped, vaporized, and reacted (combusted), creating the hot exhaust gases, which then expand through the nozzle, generating thrust. The turbopumps are fluid-flow machinery, which deliver the propellants from the tanks to the combustion chamber at high pressure and sufficient flow rate. (In some liquid-propellant rockets, the turbopumps are eliminated by using an "over-pressure" in the propellant tanks to force the propellants into the combustion chamber.) A simplified liquid-propellant rocket engine is illustrated in the figure.

Propellant tanks store one or two propellants until needed in the combustion chamber. Depending on the type of liquid propellants used, the tank may be nothing more than a low-pressure envelope, or it may be a pressure vessel capable of containing propellants under high pressure. In the case of cryogenic (extremely low temperature) propellants, the tank has to be an extremely well-insulated structure to prevent the very cold liquids from boiling away.

As with all rocket vehicle components, the mass of the propellant tanks is an important design factor. Aerospace engineers fully recognize that the lighter they can make the propellant tanks, the more payload the rocket can carry or the greater its range. Many liquid-propellant tanks are made out of very thin metal or are thin metal sheaths wrapped with high-strength fibers and cements. These tanks are stabilized by the internal pressure of their contents, much the same way a balloon's wall gains strength from the gas inside (at least up to a certain level of internal pressure). However, very large propellant tanks and tanks that contain cryogenic propellants require additional strengthening or layers. Structural rings and ribs are used to strengthen tank walls, giving the tanks the appearance of an aircraft frame. With cryogenic propellants, extensive insulation is needed to keep the propellants in their liquefied form. Unfortunately, even with the best available insulation, cryogenic propellants are difficult to store for long periods of time and eventually will boil off (i.e., vaporize). For this reason, cryogenic propellants are not used in liquid-propellant military rockets, which must be stored for months at a time in a launch-ready condition.

Turbopumps provide the required flow of propellants from the low-pressure propellant tanks to the high-pressure combustion chamber. Power to operate the turbopumps is often produced by combusting a fraction of the propellants in a preburner. Expanding gases from the burning propellants drive one or more turbines, which, in turn, drive the turbopumps. After passing through the turbines, these exhaust gases are either directed out of the rocket through a nozzle or are injected, along with liquid oxygen, into the combustion chamber for more complete burning.

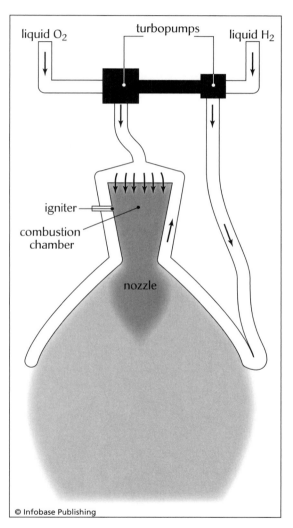

The combustion chamber of a liquid-propellant rocket is a bottle-shaped container with openings at opposite ends. The openings at the top inject propellants into the chamber. Each opening consists of a small nozzle that injects either fuel or oxidizer. The main purpose of the injectors is to mix the propellants to ensure smooth and complete combustion and to avoid detonations. Combustion chamber injectors come in many designs, and one liquid-propellant engine may have hundreds of injectors.

After the propellants have entered the combustion chamber, they must be ignited. Hypergolic propellant combinations ignite on contact, but other propellants need a starter device, such as a spark plug. Once combustion has started, the thermal energy released continues the process.

The opening at the opposite (lower) end of the combustion chamber is the throat or narrowest part of the nozzle. Combustion of the propellants builds up gas pressure inside the chamber, which then exhausts through this nozzle. By the time the gas leaves the exit cone (widest part of the nozzle), it achieves supersonic velocity and imparts forward

A simplified liquid-propellant rocket engine that employs regenerative cooling of the combustion chamber and nozzle walls thrust to the rocket vehicle. The figure shows the basic types of nozzles used in liquid-propellant rockets.

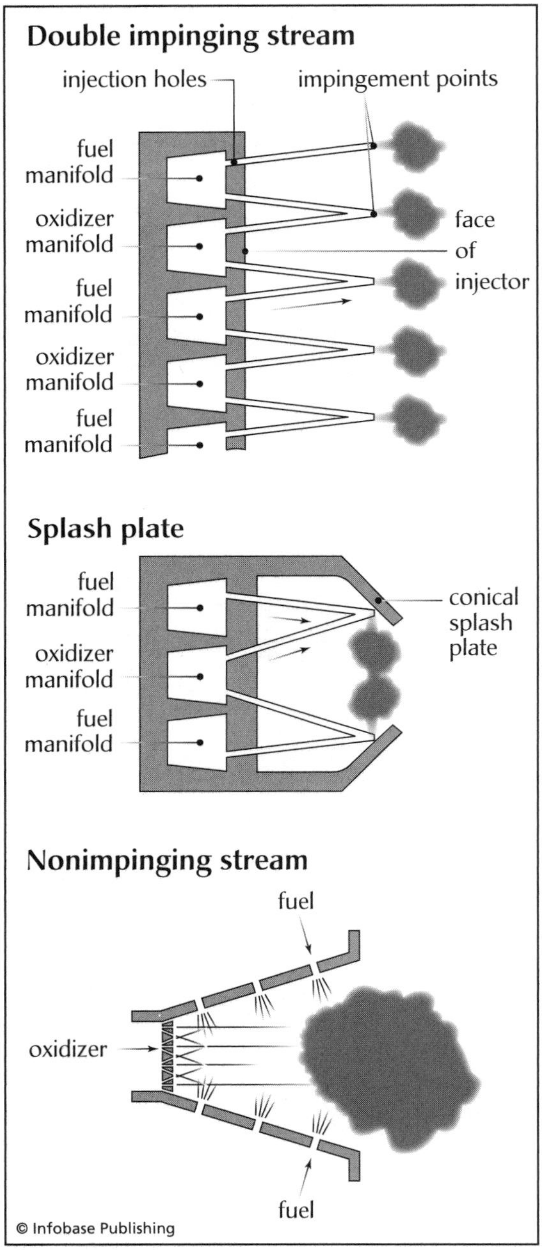

Because of the high temperatures produced by propellant combustion, the chamber and nozzle must be cooled. For example, the combustion chamber of the space shuttle's main engine (the SSME) reaches 3,590 kelvins (K)—that is, 6,000°F [3,317°C]—during firing. All surfaces of the combustion chamber and nozzle need to be protected from the eroding effects of the high-temperature, high-pressure gases.

Two general approaches can be taken to cool the combustion chamber and nozzle. One approach is identical to the cooling approach taken with many solid-propellant rocket nozzles. The surface of the nozzle is covered with an ablative material that sacrificially erodes when exposed to the high-temperature gas stream. This intentional material erosion process keeps the surface underneath cool, since the ablated material carries away a large amount of thermal energy. However, this cooling approach adds extra mass to a liquid-propellant engine, which in turn reduces payload and range capability of the rocket vehicle. Therefore, ablative cooling is used only when the liquid-propellant engine is small or when a simplified engine design is more important than high performance. In considering such technical choices, an aerospace engineer is making what are known as design tradeoffs.

The second method of cooling is called regenerative cooling. A complex plumbing arrangement inside the combustion chamber and nozzle walls circulates the fuel—in the case of the SSME, very cold (cryogenic) liquid-hydrogen fuel—before it is sent through the preburner and into the combustion chamber. This circulating fuel then absorbs some of the thermal energy entering the combustion chamber and nozzle walls, providing a level of cooling. Although more complicated than ablative cooling, regenerative cooling reduces the mass of large rocket engines and improves flight performance.

Propellants for liquid rockets generally fall into two categories: monopropellants and bipropellants.

Some typical injector designs for liquid-propellant rocket engines

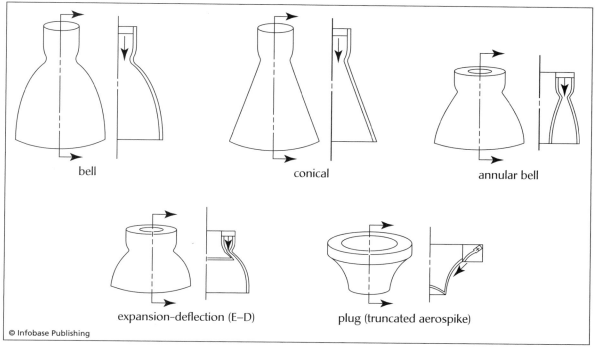

Basic types of nozzles used in liquid-propellant rocket engines

Monopropellants consist of a fuel and an oxidizing agent stored in one container. They can be two premixed chemicals, such as alcohol and hydrogen peroxide, or a homogeneous chemical, such as nitromethane. Another chemical, hydrazine, becomes a monopropellant when first brought into contact with a catalyst. The catalyst initiates a reaction that produces heat and gases from the chemical decomposition of the hydrazine.

Bipropellants have the fuel and oxidizer separate from each other until they are mixed in the combustion chamber. Commonly used bipropellant combinations include: liquid oxygen (LO_2) and kerosene, liquid oxygen (LO_2) and liquid hydrogen (LH_2), and monomethylhydrazine (MMH) and nitrogen tetroxide (N_2O_4). The last bipropellant combination, MMH and N_2O_4, is hypergolic—meaning these two propellants ignite spontaneously when brought into contact with each other. Hypergolic propellants are especially useful for attitude-control rockets where frequent firings and high reliability are required.

Aerospace engineers must consider many factors in selecting bipropellant combinations for a particular rocket system. For example, LH_2 and N_2O_4 would make a good combination based on propellant performance, but their widely divergent storage temperatures (cryogenic and room

A space shuttle main engine undergoing a full-power test firing at NASA's Stennis Space Center in Mississippi, circa May 1981. During this particular development test, the powerful and complex liquid-propellant engine operated at full power for 290 seconds. *(NASA)*

temperature, respectively) would require the use of large quantities of thermal insulation between the two tanks, adding considerable mass to the rocket vehicle. Another important factor is the toxicity of the chemicals used. MMH and N_2O_4 are both highly toxic. Rocket vehicles that use this propellant combination require special propellant handling and prelaunch preparation.

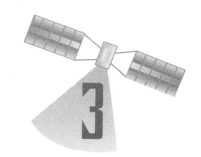

Pioneering Rocket Planes

This chapter focuses on the interesting topic of rocket planes—aircraft that use a rocket engine for propulsion. In the early through mid-20th century, aviation and rocket enthusiasts followed in the footsteps of the legendary Wan-Hu, as they pioneered novel combinations of the rocket engine with various types of airplanes and gliders. Their efforts not only blazed trails in aviation but also led human beings to the threshold of space.

In the American aerospace program, an experimental aircraft, missile, or aerospace vehicle is often designated by the symbol X. Two of the most interesting and historically significant "X-planes" were the X-1 and X-15 rocket-powered research aircraft. The early rocket planes generally achieved much higher speeds that similarly sized jet-propelled aircraft. However, because of propellant supply limits, they could only operate for only short periods of time. The completely reusable aerospace vehicle, or spaceplane, now represents the Holy Grail of 21st-century space technology.

✧ Rocket–Assisted Takeoff

In the mid-1930s, professor Theodore von Kármán, head of the Guggenheim Aeronautical Laboratory at the California Institute of Technology (GALCIT), initiated pioneering work in rocket propulsion that would have profound impact on aeronautics and astronautics. Several of his graduate students and technical assistants gathered to test a primitive rocket engine in a dry riverbed wilderness area in the Arroyo Seco—a dry canyon wash north of the Rose Bowl in Pasadena, California. On October 31, 1936, the group conducted their first rocket firing at this location.

Von Kármán also served as an advisor to the U.S. Army Air Corps. So after the Caltech group's successful rocket experiment, he persuaded the U.S. Army to fund development of strap-on rockets to help overloaded military aircraft take off from short runways. This use of rockets to help launch aircraft was initially called "rocket-assisted takeoff," or RATO. However, during World War II, the expression "jet-assisted takeoff," or JATO, came into popular use and now is the more commonly recognized term.

In support of von Kármán's rocket-assisted takeoff project, the U.S. Army helped Caltech acquire land for test pits and temporary workshops. Airplane tests at nearby air bases soon proved the validity of the concept

Takeoff of the first American "rocket-assisted" airplane—an Ercoupe fitted with a solid-propellant rocket booster developed by Theodore von Kármán's group at Caltech. Piloted by Captain Homer A. Boushey, Jr., of the U.S. Army Air Corps, this aircraft took off from March Field on August 12, 1941. The successful demonstration of the rocket-assisted takeoff (RATO) technique gave rise to its use by certain types of military aircraft during and after World War II. Such rocket boosters became more commonly known as jet-assisted takeoff (JATO) units. *(NASA)*

THEODORE VON KÁRMÁN
(1881-1963)

The Hungarian–American mathematician and research engineer Theodore von Kármán cofounded the Jet Propulsion Laboratory (JPL) Pasadena, California. As an aeronautical theoretician, he made numerous contributions to the fields of aerodynamics and aeronautical engineering. In addition to being cofounder of JPL, he was also the principal founder of a major rocket propulsion firm (Aerojet-General Corporation), the top science advisor to the U.S. Air Force during its transition to jet propulsion aircraft, and the top science advisor to the North Atlantic Treaty Organization (NATO). During much of this time, he served as the fountainhead of aerodynamic thought in his capacity as director of the Guggenheim Aeronautical Laboratory at the California Institute of Technology (GALCIT) in Pasadena, California.

Under his guidance, Caltech's 10-foot (3-m) wind tunnel was designed, built, and operated. Many industrial firms tested new aeronautical designs and concepts in GALCIT's wind tunnel. From 1936 to 1940, Caltech stood alone as the only university-based rocket research center. Von Kármán gambled his academic and professional prestige by supporting work on rocketry. Other institutions of higher learning dismissed such research as "fantastical" and left such endeavors to visionaries like Robert H. Goddard. Important theoretical research by von Kármán gave rise to the first successful solid-fuel rocket engine firings at Caltech. This effort led to federal funding for studies, which ultimately resulted in a widely used form of aircraft rocket propulsion called jet-assisted takeoff (JATO). Success in the JATO endeavor encouraged von Kármán to establish two more highly regarded institutions, both originally dedicated to rocketry: the Aerojet Engineering Company and the Jet Propulsion Laboratory.

The last years of his life were spent in Paris, his favorite city. However, his interest in aeronautical research and his contributions to the field never waned. In Paris, he organized the NATO Advisory Group for Aeronautical Research and Development (AGARD). Staffed by American and European scientists, its many committees investigated such disciplines as propulsion, aerodynamics, and electronics. The legacy of his personable leadership and "soft touch" approach to problem solving was equaled only by his genius. In 1963, President John F. Kennedy awarded von Kármán the first National Medal of Science.

and allowed von Kármán's team to test and improve various designs. The first successful rocket-assisted airplane takeoff in the United States took place on August 12, 1941, at March Air Force Base in California. The pilot was Captain (later Brigadier General) Homer A. Boushey, Jr., of the United States Army Air Corps. He flew a small civilian-type airplane (called the Ercoupe) that belonged to the U.S. Army. Von Kármán's group had fitted this plane with a solid-propellant rocket booster, which had a thrust of 28 pounds-force (125 N). Boushey's plane took off in less than half the normal runway distance used by the Ercoupe—effectively demonstrating a valuable application of rocket propulsion that continues to the present day. The arrival of World War II placed a great demand for von Kármán's JATO units, first from the U.S. Army Air Corps and then from various aviation components of the U.S. Navy.

✦ *Messerschmitt 163*

During World War II, the German Air Force (Luftwaffe) developed and deployed the world's only tailless operational rocket-powered fighter aircraft. Called the *Messerschmitt 163* (Me-163), or *Komet,* this unusual military aircraft experienced its first test flight in the spring of 1941. Then, after a rather delayed development, the Me-163 entered service in 1944 as an interceptor—primarily in a point defense role because of its short range and limited operating period.

The Me-163 was a stubby-looking, swept-winged rocket-powered military aircraft. The operational version of the rocket plane was called the ME-163 B-1a. This plane had a length of almost 20 feet (6 m), a height of 10 feet (3 m), and a wingspan of almost 31 feet (9.5 m). The Komet carried a crew of one (the pilot) and could operate for only about eight minutes. Operation was limited by the rapid propellant consumption of its single Walter HWK 509A-2 rocket engine, which produced a thrust of 3,825 pounds-force (17,000 N). This limited period of operation proved quite disadvantageous in combat. The Me-163 would rise at an astounding rate to intercept Allied bombers, but then would zip right past the lumbering bombers with little opportunity to fire its two 30-mm cannons. After one or (possibly) two passes, the Me-163 would run out of fuel and have to glide back to its air field, where it would land on an extended skid. Allied fighter aircraft could not match the speed of the Me-163, but American and British fighter pilots soon developed effective countermeasures that exploited the *Komet*'s weaknesses. The Allied fighter pilots would exercise patience and then follow the fuel-expended Komet back to its airfield— and destroy the German rocket plane on the ground. As a result of these limitations, more Komets were destroyed in combat than destroyed Allied aircraft.

The fuel used by this German rocket plane also proved extremely hazardous to the pilot. The Komet's rocket engine burned a volatile mixture of "T-Stoff" (a mixture containing 80 percent hydrogen peroxide and 20 percent water) and "C-Stoff" (a mixture of methyl alcohol, hydrazine hydrate, and water). The rocket plane's engine would sometimes experience serious cavitation problems, leading to a catastrophic explosion. Other times, the troublesome landing skid would fail to deploy properly, causing the Me-

163 to flip over during its high speed, but unpowered, glide landing. In these cases, the leakage of any remaining fuel often caused an explosion or gave the pilot severe chemical burns.

The Me-163 was a technically interesting rocket plane that saw only limited duty as an interceptor. Like other wonder weapons (*Wunderwaffen*), the Komet arrived too late to prevent the defeat of Nazi Germany at the end of World War II. By the end of World War II, only 279 operational rocket planes were delivered to the German Air Force. In fact, there was only one operational Komet combat group formed (called JG 400)—it destroyed nine Allied aircraft, while losing 14 of its own. The Me-163 could climb to its service ceiling, an altitude of about 39,000 feet (12,000 m) in about 200 seconds, but only carried enough fuel to operate in powered flight for 480 seconds (about eight minutes). Despite its operational shortcomings, the rocket plane achieved important technical milestones. Most notably, with a maximum speed of about 596 miles per hour (960 km/h), the Komet came very close to breaking the sound barrier and achieving supersonic flight.

Following World War II, American intelligence officers examined several captured Me-163 rocket planes to see if an advanced version of this human-piloted rocket interceptor might serve a useful defense role against enemy aircraft. However, rapid improvements in surface-to-air missile technology during the early portions of the cold war (see chapter 4) soon eliminated any further consideration of human-piloted military rocket planes in point defense roles.

✧ X–1 Rocket–Powered Research Airplane

The rocket-powered Bell X-1 research aircraft—patterned on the lines of a 50-caliber machine-gun bullet—was the first human-crewed vehicle to fly faster than the speed of sound. The speed of sound in air varies with altitude. On October 14, 1947 the Bell X-1, named "Glamorous Glennis" and piloted by Captain Charles "Chuck" Yeager, was carried aloft under the bomb bay of a B-29 bomber and then released. The pilot ignited the aircraft's rocket engine, climbed, and accelerated, reaching Mach 1.06, or 700 miles per hour (1,127 km/h) as it flew over Edwards Air Force Base in California at an altitude of 43,000 feet (13,100 m). At this altitude, the speed of sound (or Mach 1.00) is 670 miles per hour (1,078.7 km/h). The rocket-powered experimental aircraft, having used up all of its propellant, then glided to a landing on its tricycle gear at Muroc Dry Lake in the Mojave Desert.

Bell Aircraft developed the X-1 largely based on design criteria provided by National Advisory Committee for Aeronautics (NACA) and

Piloted by U.S. Air Force Captain Charles Yeager, this Bell X-1 rocket-propelled research aircraft broke the sound barrier on October 14, 1947, in the sky over Edwards Air Force Base in California—a flight-test facility formerly called Muroc Army Air Field. *(NASA)*

the U.S. Army Air Corps. The primary objective of this project and this research aircraft was transonic flight. Aeronautical engineers recognized that rocket-plane technology was preferred to the jet-plane technology available in the mid-1940s in achieving this important technical milestone. The X-1 research aircraft measured 30.8 feet (9.4 m) in length and 10.8 feet (3.30 m) in height and has a wingspan of 27.9 feet (8.5 m). The X-1 had a total gross mass of 6,790 pounds-mass (3,085 kg) (empty) and 13,035 pounds-mass (5,925 kg) (loaded).

The X-1 planes were equipped with an array of internal NACA flight data recorders. Pilot instrumentation was conventional with the exception of the Machmeter, an adjustable stabilizer switch, and the rocket engine controls. The cockpit had room for one person, the pilot. It was pressurized to a maximum of 3.0 psi (20.7 kilopascals) by nitrogen gas. An H-shaped control wheel rather than a conventional stick was provided to allow the pilot to use both hands in controlling X-1 flight in the expected turbulence of the transonic range.

The X-1's 211 pound-mass (96-kg) rocket motor (a Reaction Motors XLR-11 engine) was a four-chamber unit providing a nominal 3,820 pounds-force (17,000 N) of thrust at sea level. The engine could not be throttled, although each chamber could be fired individually. The on-board propellant supply for the rocket engine used on the first two X-1 aircraft consisted of 311 gallons (1,180 L) of liquid oxygen and 293 gallons (1,110 L) of diluted ethyl alcohol. The propellant capacity for the third X-1 aircraft was 437 gallons (1,655 L) of liquid oxygen and 492 gallons (1,865 L) of alcohol.

The first two X-1 aircraft used a high-pressure nitrogen gas system to force the liquid oxygen and alcohol into the rocket engine. The 17.5 cubic foot (0.5 m^3) supply of nitrogen was stored inside the airplane in twelve pressurized tanks. Due to the mass of the nitrogen tanks, the fuel supply for the first two X-1 aircraft constructed was severely restricted. Although these two research aircraft were initially designed to carry 8,160 pounds-mass (3,710 kg) of propellant, the changes required by the nitrogen tankage dropped propellant capacity to only 5,038 pounds-mass (2,290 kg). This small propellant supply limited the first two X-1 rocket planes to a maximum of about 2.5 minutes of powered flight time. The restricted flight time severely hampered efforts to reach the desired 35,100 foot (10,700 m) altitude, where supersonic testing was planned. This circumstance led to the decision to use an air launch system for the X-1.

Bell Aircraft engineers initially selected a Boeing B-29 mother ship to carry the X-1 aloft to an altitude of approximately 25,000 feet (7,625 m) for air launch powered-flight testing. Their decision to employ a mother ship to carry the X-1 to launch altitude was strictly based on the X-1's limited operating time. The third X-1 research aircraft delivered by Bell Aircraft used a B-50A mother ship for its flight. On occasion, B-50 aircraft were also used to launch the first and second X-1 aircraft delivered to the U.S. Army Air Corps, which funded the overall project.

No ejection seat was included in the construction of the X-1 because of mass involved and questions of its utility in high-speed escape. Nor was there any proven method of exit through the cockpit hatch, due to the close proximity to the wing leading edge and the necessity to unhinge the pilot control wheel. Data from the X-1 flight research validated (and in some cases, revealed the limitations of) transonic wind tunnel data and theoretical analysis developed elsewhere. The research techniques used in the X-1 program became the pattern for all subsequent X-craft projects. The results of X-1 test series also helped lay the foundation of America's aerospace program in the 1960s and beyond.

✧ X-15 Rocket Plane

The North American X-15 was a rocket-powered experimental aircraft, which was 50 feet (15.24m) long and had a wingspan of 22 feet (6.71 m). This rocket plane was a missile-shaped vehicle with an unusual wedge-shaped vertical tail, thin stubby wings, and unique fairings that extended along the side of the fuselage. The X-15 had an empty mass of 13,950 pounds (6,340 kg) and a launch mass of 33,925 pounds (15,420 kg). The vehicle's pilot-controlled rocket engine was capable of developing 57,000 pounds-force (253,500 N) of thrust at sea level and 70,000 pounds-force (311,000 N) at peak altitude. The X-15's rocket engine used anhydrous ammonia as its fuel and liquid oxygen as the oxidizer.

The X-15 research aircraft helped bridge the gap between human flight within the atmosphere and human flight in space. The vehicle was developed and flown in the 1960s to provide in-flight information and data on aerodynamics, structures, flight controls, and the physiological aspects of high-speed, high-altitude flight. For flight in the dense air of the lower ("aircraft-usable") portions of the atmosphere, the X-15 employed conventional aerodynamic controls. However, for flight in the thin upper portions of Earth's atmospheric envelope, the X-15 used a ballistic control system. Eight hydrogen peroxide-fueled thruster rockets, located on the nose of the aircraft, provided pitch and yaw control.

Because of its large fuel consumption, the X-15 was launched from a B-52 aircraft (i.e., a "mother ship") at an altitude of about 45,000 feet (13,700 m) and a speed of about 500 miles per hour (805 km/h). Then the

An X-15 rocket plane launches away from its B-52 mother ship over Edwards Air Force Base, California, in the early 1960s. Seconds after air launch, the pilot ignites the X-15's rocket engine (as shown here). The white patches appearing near the middle of the aircraft are frost from the liquid oxygen used in the rocket plane's propulsion system. *(NASA)*

NASA pilot Neil Armstrong is seen here next to the X-15 rocket plane after a successful hypersonic research flight (1960). On July 20, 1969, as an *Apollo 11* mission astronaut, Armstrong became the first human being to walk on the surface of the Moon. *(NASA)*

pilot ignited the rocket engine, which provided thrust for the first 80 to 120 seconds of flight, depending on the type of mission being flown. The remainder of the normal 10- to 11-minute duration flight was powerless and ended with a 200-mile-per-hour (322-km/h) glide landing at Edwards Air Force Base in California. Generally, one of two types of X-15 flight profiles was used: a high-altitude flight plan that called for the pilot to maintain a steep rate of climb; or a speed profile that called for the pilot to push over and maintain a level altitude.

First flown in 1959, the three X-15 aircraft made a total of 199 flights. The X-15 flew more than six times the speed of sound and reached a maximum altitude of 67 miles (107.9 km) and a maximum speed of 4,517 miles per hour (7,273 km/h). On August 22, 1963, the X-15 set an altitude record—67 miles [107.9 km]—during a flight that took the pilot (Joseph

A. Walker) to the edge of outer space. The final X-15 flight occurred on October 24, 1968. It is interesting to note that Apollo astronaut Neil Armstrong (the first human to walk on the Moon) was one of the pilots who flew the X-15 aircraft.

This joint program by NASA, the U.S. Air Force, the U.S. Navy, and North American Aviation operated the most remarkable of all the American rocket-powered research aircraft. The X-15 rocket plane provided an enormous wealth of data on hypersonic air flow, aerodynamic heating, control and stability at hypersonic speeds, reaction controls for flight above the atmosphere, piloting techniques for reentry, human factors, and flight instrumentation. The highly successful X-15 program contributed to the development of the NASA's Mercury, Gemini, and Apollo piloted spaceflight projects, as well as the space shuttle program.

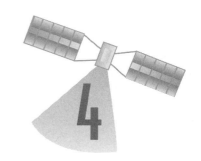

Military Missiles and Rockets

Beginning in World War II and continuing throughout the cold war, a variety of guided missiles—both jet propelled and rocket propelled—emerged to support a wide range of tactical and strategic military applications. Stimulated by advances in nuclear weapons technology, the miniaturization of electronics, and the development of ever more powerful computers, these missiles and rockets have now become an integral part in modern warfare. This chapter discusses the major types of military missiles and rockets and includes several examples of historically important or currently operational systems.

✧ Guided Missiles

Military technologists define the guided missile as an unmanned, self-propelled vehicle moving above the surface of Earth whose trajectory or course is capable of being controlled while in flight. An air-to-air missile (AAM) is an air-launched vehicle for use against aerial targets. An air-to-ground missile (AGM) is an air-launched missile for use against surface targets, both on land and at sea. A surface-to-air missile (SAM) is a surface-launched guided missile for use against targets in the air, including enemy aircraft and hostile cruise missiles and ballistic missiles. Finally, a surface-to-surface missile (SSM) is a surface-launched missile for use against surface targets, either tactical or strategic. Tactical targets include enemy tanks, fortified positions, troop concentrations, communications centers, and similar battlefield-related items. Strategic targets include war-making facilities deep in an enemy's territories, such as naval installations, airfields, transportation and communications networks, and industrial centers.

An early test flight (July 18, 1951) of the United States Air Force's Matador surface-to-surface tactical missile at Cape Canaveral Air Force Station, Florida. Designated the SSM-A-1, the Matador was the first operational guided missile of the U.S. armed forces. It was launched by a booster rocket from a mobile trailer and controlled electronically from the ground during flight. Immediately after launch, the booster rocket fell away and a 254,000-newton thrust jet engine powered the guided missile to its intended target. *(United States Air Force)*

Modern missiles have been designed to carry a range of conventional (high-explosive) warheads, as well as weapons of mass destruction—that is, nuclear, chemical, or biological weapon payloads. Current treaties and international conventions prohibit signatory nations (including the United States) from using missile-delivered chemical or biological weapons.

The air-breathing missile is a guided missile with an engine requiring the intake of air for combustion of its fuel, such as a ramjet or turbojet. It can be contrasted with the rocket-engine missile, which carries its own oxidizer and can operate beyond the atmosphere.

✦ Tactical Missiles Come in All Sizes

AIR-TO-AIR MISSILE

The air-to-air missile (AAM) is a missile launched from an airborne vehicle at a target above the surface of Earth (that is, at a target in the air). One example is the Advance Medium Range, Air-to-Air Missile, or AMRAAM. Designated as the AIM-120, this missile was developed by the United States Air Force and Navy in conjunction with several NATO allies. AMRAAM represents the next-generation follow-on to the Sparrow (AIM-7) air-to-air missile. The AIM-120 missile is faster, smaller, and lighter and has improved capabilities against low-altitude targets. Powered to supersonic speeds by a solid-propellant rocket, the 12-foot- (3.7-m-) long AMRAAM currently serves the U.S. Air Force, the U.S. Navy, and allies of the United States as an all-weather, medium range, air-to-air tactical missile.

AIR-TO-GROUND MISSILE

A wide variety of air-to-surface guided missiles have been developed to enable military aircraft to perform precision attacks on surface targets from relatively safe standoff distances. One example of this type of air-to-ground (AGM) missile is the AGM-130A missile, equipped with a television and imaging infrared seeker that allows an airborne weapon system officer to observe and/or steer the missile as it attacks a ground target. Developed by the United States Air Force, the AGM-130A supports high- and low-altitude precision strikes at standoff ranges.

The U.S. Army's Honest John missile being launched from a standard military truck. Starting in the early 1950s, the U.S. Army used this unguided rocket as a substitute for long-range artillery in providing close support of front-line troops. The Honest John employed a solid-propellant rocket motor, had a maximum range of 19 km, and could carry either a nuclear or conventional (high-explosive) warhead. (*U.S. Army/White Sands Missile Range*)

Another example is the Maverick (AGM-65)—a tactical guided missile designed for close air support, interdiction, and defense suppression missions. Powered by a solid-propellant rocket motor, the 8.2-foot-long (2.5-m) Maverick has a range in excess of 17 miles (27 km). It provides standoff capability and high-probability strike against a wide range of tactical targets, including armor, air defenses, ships, transportation equipment, and fuel storage facilities. First entering operational service in the 1970s, evolved versions of the Maverick missile are now being used by aircraft in the U.S. Air Force, Navy, and Marine Corps.

Another interesting American air-to-surface missile is called the HARM missile. The AGM-88 High-Speed Antiradiation Missile (HARM) is an air-to-surface tactical missile designed to seek out and destroy enemy radar-equipped air defense systems. An antiradiation missile passively homes in on enemy electronic signal emitters, especially those associated with radar sites used to direct antiaircraft guns and surface-to-air missiles.

The AGM-88 can detect, attack, and destroy a target with minimum aircrew input. The U.S. Air Force and Navy use this missile extensively.

The Hellfire (AGM-114) is a family of air-to-ground missiles developed by the U.S. Army to provide heavy antiarmor capability for attack helicopters. The first three generations of Hellfire missiles used a laser seeker, while the fourth-generation missile, Longbow Hellfire, employs a radio frequency or radar seeker. As part of the war on terrorism, Hellfire missiles have been carried by and successfully fired from the Predator unmanned aerial vehicle. The U.S. Navy and Marine Corps use versions of the Hellfire missile (namely, the AGM-114B/K/M models) as an air-to-air weapon against hostile helicopters or slow-moving fixed-wing aircraft.

Finally, the Harpoon missile (AGM-84D) is an all-weather, over-the-horizon, antiship missile. The Harpoon's active radar guidance, warhead design, and low-level, sea-skimming cruise trajectory assure high survivability and effectiveness. This missile is capable of being launched from surface ships, submarines, or (without its booster component) from aircraft. Originally developed for the U.S. Navy to serve as its basic antiship missile for fleetwide use, the AGM-84D also has been adapted for use on the U.S. Air Force's B-52G bombers.

SURFACE-TO-AIR MISSILE

The surface-to-air missile, or SAM, is designed to provide point defense against attacking military aircraft, and possibly incoming ballistic missiles or cruise missile. One of the most well known surface-to-air missiles is the U.S. Army's Stinger missile. The shoulder-fired Stinger weapon system provides effective, short-range air-defense capabilities to U.S. Army soldiers and U.S. Marine Corps (USMC), U.S. Navy (USN), and U.S. Air Force (USAF) ground personnel who defend important sites against low-level fixed and rotary-wing aircraft attack. It is the weapon of choice for the United States and many allied armed forces. The 35 pound-mass (16 kg), supersonic, fire-and-forget Stinger features quick-reaction acquisition and tracking, and all-aspect engagement, including head-on. Improved missile speed, range, maneuverability, flight tracking, and countermeasures rejection made the individual combat air defender using the Stinger the equal of the most sophisticated threat aircraft.

Three variants, Basic Stinger, Stinger–Passive Optical Seeker Technique (POST), and Stinger–Reprogrammable Microprocessor (RMP) were developed. All operate in a similar fashion; are fielded as certified rounds in a disposable, sealed launch tube; and require no field maintenance. All three use the rolling airframe concept, proportional navigation, passive homing, separate launch motor, dual-thrust flight motor, penetrating hit-to-kill warhead, reusable launcher/gripstock, and belt-pack identification, friend or foe (IFF) transceiver.

U.S. Army soldier test firing a Stinger missile at the White Sands Missile Range in New Mexico. The Stinger is a supersonic, fire-and-forget air-defense weapon against low-level fixed-wing and rotary-wing aircraft attack. *(U.S. Army/White Sands Missile Range)*

The basic Stinger is an infrared reticle-scan analog system using discrete component signal processing. The Stinger-POST employs an infrared/ultraviolet dual detector, rosette-pattern image scanning, and digital microprocessor-based signal processing. Advanced features include improved acquisition, false target rejection, and additional countermeasures capabilities. The Stinger-RMP adds additional microprocessor power and is highly countermeasure-resistant. New external software reprogrammability allows upgrades without costly retrofit as the threat evolves. The U.S. Army uses the

Stinger-RMP not only for ground forces but to arm helicopters for air-to-air combat (Air-to-Air Stinger) and on the Bradley Fighting Vehicle (BFV) to provide low-altitude air defense. Many allied countries have selected the export version of the Stinger-RMP for their armed forces.

ANTIMISSILE–MISSILE

An antimissile-missile (AMM) is a missile launched against a hostile missile in flight. However, stopping a ballistic missile in flight is a very difficult task, especially in the terminal phase of its trajectory, when the task has been likened to hitting one bullet with another bullet.

The Sprint was a high-acceleration, short-range, nuclear-armed, surface-to-air guided missile that was deployed by the U.S. Army in 1975 as part of the now-defunct Safeguard ballistic missile defense system. The nuclear-armed Sprint system was designed to intercept strategic ballistic reentry vehicles in the atmosphere during the terminal phase of their flight.

An early United States antimissile-missile, called the Sprint, being tested at the White Sands Missile Range in 1967 *(U.S. Army/White Sands Missile Range)*

Today, the U.S. Army's Patriot missile system provides high- and medium-altitude defense against aircraft and tactical ballistic missiles. The combat element of the Patriot missile system is the fire unit that consists of a radar set, an engagement control station (ECS), an equipment power plant (EPP), an antenna mast group (AMG), and eight remotely located launchers. The single phased-array radar provides the following tactical functions: airspace surveillance, target detection and tracking, and missile guidance. The ECS provides the human interface for command and control operations. Each firing battery launcher contains four ready-to-fire missiles sealed in canisters that serve a dual purpose as shipping containers and launch tubes. The Patriot's fast reaction capability, high firepower, ability to track up to 50 targets simultaneously with a maximum range of 42.5 miles (68.5 km), and ability to operate in a severe electronic countermeasures (ECM) environment are features not available in previous air defense systems. The U.S. Army received the Patriot in 1985, and the system gained notoriety during the Persian Gulf War of 1991 as the "Scud killer."

The Missile Defense Agency (MDA) within the Department of Defense is currently managing the Patriot Advanced Capability-3 (PAC-3) upgrade program. PAC-3 is a terminal defense system being built upon the previous Patriot and missile defense infrastructure. The PAC-3 missile is a high-velocity, hit-to-kill vehicle that represents the latest generation of Patriot missile. It is being developed to provide an increased capability against tactical ballistic missiles, cruise missiles, and hostile aircraft. The PAC-3 missile provides the range, accuracy, and lethalness to effectively defend against theater ballistic missiles (TBMs) that may be carrying either conventional high-explosive, biological, chemical, or nuclear warheads. Unlike previous Patriot missiles, the PAC-3 uses an active radar seeker and closed loop guidance to directly hit the target.

With the PAC-3 missile and ground system, threat missiles can be engaged and destroyed at higher altitudes and greater ranges with better lethalness. The system can operate despite electronic countermeasures. Due to the active seeker and closed loop guidance, a greater number of interceptors can be controlled at one time than was possible with earlier Patriot missiles.

There is a growing international threat of rogue nations acquiring theater ballistic missiles for use against neighboring nations. One example of immediate concern is that of North Korea threatening both South Korea and Japan with nuclear-armed ballistic missiles. The Missile Defense Agency has the mission to develop, test, and prepare for deployment an American missile defense system. It is integrating advanced interceptors, land-, sea-, air- and space-based sensors, and battle management command and control systems in order to design and demonstrate a layered missile defense system that can respond to and engage all classes and ranges of ballistic missile threats. The contemporary missile defense systems being developed and tested by the agency are primarily based on hit-to-kill vehicle technology and sophisticated, data-fusing battle management systems.

✧ Cruise Missile

A cruise missile is a guided missile traveling within the atmosphere at aircraft speeds and, usually, low altitude whose trajectory is preprogrammed. It is capable of achieving high accuracy in striking a distant target. It is maneuverable during flight and, because it is constantly propelled, does not follow a ballistic trajectory. Cruise missiles may be armed with nuclear weapons or with conventional warheads (i.e., high explosives).

The Tomahawk is a long-range, subsonic cruise missile used by the U.S. Navy for land attack and for antisurface warfare. Tomahawk is an all-weather submarine or ship-launched antiship or land-attack cruise missile. After launch, a solid-propellant rocket engine propels the missile until a small turbofan engine takes over for the cruise portion of the

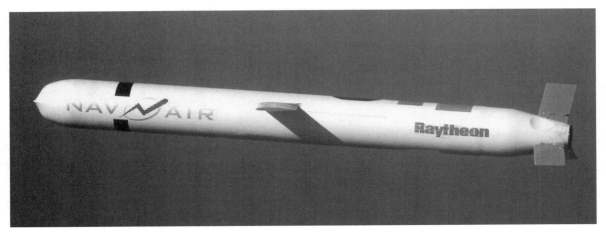

A U.S. Navy Tactical Tomahawk Block IV cruise missile conducts a controlled flight test over the Naval Air Systems Command (NAVAIR) western test-range complex in southern California on November 10, 2002. During this flight test, the cruise missile successfully completed a vertical underwater launch, flew a fully guided 780-mile (1,255-km) course, and impacted a designated target structure as planned. The Tactical Tomahawk is the next generation of Tomahawk cruise missile with many information age improvements—including the capability to reprogram this missile while it is in flight to strike alternate targets. *(U.S. Navy)*

flight. This cruise missile is a highly survivable weapon. Radar detection is difficult because of its small cross section and low-altitude flight. Similarly, infrared detection is also difficult because the turbofan emits little heat.

The antiship variant of Tomahawk uses a combined active radar seeker and passive system to seek out, engage, and destroy a hostile ship at long range. Its modified Harpoon cruise missile guidance system permits the Tomahawk to be launched and fly at low altitudes in the general direction of an enemy warship to avoid radar detection. Then, at a programmed distance, the missile begins an active radar search to seek out, acquire, and hit the target ship.

The land-attack version has inertial and terrain contour matching (TERCOM) guidance. This guidance system uses a stored map reference to compare with the actual terrain to help the missile determine its position. If necessary, a course correction is made to place the missile on course to the target.

The basic Tomahawk is 18.2 feet (5.56 m) long and has a mass of 2,622 pounds (1,192 kg), not including the booster. It has a diameter of 20.4 inches (51.81 cm) and a wingspan (when deployed) of 8.75 feet (2.67 m). The power plant for this cruise missile includes a turbofan jet engine and a solid-fuel booster rocket. This missile is subsonic and cruises at about 550 miles per hour (880 km/h). It can carry a conventional or nuclear warhead. In the land-attack (conventional warhead) configuration, it has a range of 685 miles (1,100 km), while in the land-attack (nuclear warhead) configuration, it has a range of 1,540 miles (2,480 km). In the antiship role, it has a range of over 285 miles (460 km). This missile was first deployed in 1983.

✦ Ballistic Missile

A ballistic missile is a missile that is propelled by rocket engines and guided only during the initial (thrust producing) phase of its flight. In the nonpowered and nonguided phase of its flight, it assumes a ballistic trajectory similar to that of an artillery shell. The ballistic trajectory is the path an object (that does not have lifting surfaces) follows while being acted upon only by the force of gravity and any resistive aerodynamic forces of the medium through which it passes. A stone tossed into the air follows a ballistic trajectory. Similarly, after its propulsive unit stops operating, a rocket vehicle describes a ballistic trajectory.

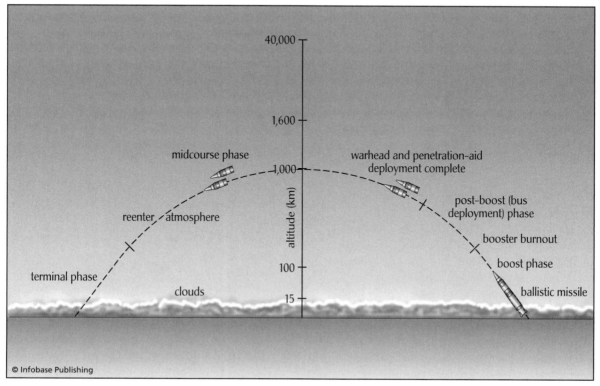

Major phases of a typical ballistic missile flight trajectory

The figure at the bottom of page 67 shows the basic phases of the flight of a long-range ballistic missile, called an intercontinental ballistic missile (ICBM). The boost phase is that portion of the flight during which the booster engine(s) and sustainer engine(s) operate. In the first portion of this trajectory, the ballistic missile is being powered by its rocket engines. During this period, which typically lasts from three to five minutes for an ICBM, the missile reaches an altitude of about 125 miles (200 km), whereupon powered flight ends and the ICBM begins to dispense its reentry vehicles. The other portions of the missile flight, including the midcourse phase and the reentry phase, take up the remainder of an ICBM's flight total time of about 25 to 30 minutes. In the bus-deployment phase of an ICBM's flight, multiple warheads and, possibly, decoy reentry vehicles are deployed on different paths to different targets. The warheads on a single ICBM are carried on a platform, called the bus (or post-boost vehicle), which has small rocket motors to move it slightly from its original flight path. This phase is also called the post-boost phase. After thrust termination, reentry vehicles (RVs) can be released and these RVs also follow free-falling (ballistic) trajectories toward their targets.

Aerospace analysts within the U.S. Department of Defense (DOD) often classify ballistic missiles by their maximum operational ranges, using the following scale: short-range ballistic missiles (SRBMs) are those that have a maximum operational range of about 685 miles (1,100 km); medium-range ballistic missiles (MRBMs) have an operational range between 685 miles (1,100 km) and 1,710 miles (2,750 km); intermediate-range ballistic missiles (IRBMs) have an operational range between 1,710 miles (2,750 km) and 3,420 miles (5,500 km); and intercontinental ballistic missiles have operational ranges in excess of 3,420 miles (5,500 km). While somewhat arbitrary from an aerospace engineering perspective, this widely recognized classification scheme has proven quite useful in arms-control negotiations, ballistic missile treaty discussions, and international initiatives focused on limiting regional arms races and preventing the emergence of far-reaching ballistic missile threats from rogue nations.

U.S. Army personnel prepare to launch a liquid-fueled Redstone intermediate-range ballistic missile (IRBM) at the White Sands Missile Range in the early 1950s. A direct technical descendant of the German V-2 military rocket, the Redstone rocket was also pressed into service as a space launch vehicle—helping to successfully place the first American satellite (*Explorer 1*) into orbit in 1958 and then propelling the Mercury Project astronauts, Alan B. Shepard, Jr., and Virgil "Gus" I. Grissom, on the first two successful American manned space missions (both suborbital) in 1961. *(U.S. Army/White Sands Missile Range)*

Intercontinental ballistic missiles are either land-based (in fixed silos or on mobile platforms) or deployed at sea on nuclear-powered submarines.

The United States Air Force's Minuteman ICBM is a three-stage solid-propellant ballistic missile that is guided to its target by an all-inertial guidance and control system. These strategic missiles are equipped with nuclear warheads and designed for deployment in hardened and dispersed underground silos. Designated as the LGM-30, the Minuteman ICBM is an element of the U.S. strategic deterrent force. The L in LGM stands for silo-configuration; G for surface attack; and M for guided missile.

The Minuteman weapon system was conceived in the late 1950s and deployed in the mid-1960s. Minuteman was a revolutionary concept and an extraordinary technical achievement. Both the missile and basing components incorporated significant advances beyond the relatively slow-reacting, liquid-fueled, remotely controlled ICBMs of the previous generation of missiles (such as the Atlas and the Titan). From the beginning, Minuteman missiles have provided a quick-reacting, inertially guided, highly survivable component of America's nuclear Triad. Minuteman's maintenance concept capitalizes on high reliability and a "remove-and-replace" approach to achieve a near 100-percent alert rate.

Through state-of-the-art improvements, the Minuteman system has evolved over three decades to meet new challenges and assume new missions. Modernization programs have resulted in new versions of the missile, expanded targeting options, and significantly improved accuracy. For example, when the Minuteman I became operational in October 1962, it had a single-target capability. The Minuteman II became operational in October 1965. Though it looked similar to the Minuteman I, the Minuteman II had greater range and targeting capability. Finally, the Minuteman III became operational in June 1970. This missile, with its improved third stage and the postboost vehicle, can deliver multiple, independently targeted reentry vehicles and their penetration aids onto different targets. Over 500 Minuteman IIIs are currently deployed at bases in the United States.

The fleet ballistic missile (FBM) is an intercontinental ballistic missile, usually equipped with one or more nuclear weapons, carried by and launched from a submarine. The U.S. Navy equips a special class of nuclear powered submarines, called fleet ballistic missile submarines, with such nuclear warhead–carrying missiles. The Trident II (designated D-5) is the sixth-generation member of the U.S. Navy's FBM program that started in 1956. Since then the Polaris (A1), Polaris (A2), Polaris (A3), Poseidon (C3), and Trident I (C4) have provided a significant deterrent against nuclear aggression. At present, the U.S. Navy deploys Poseidon (C3) and Trident I (C4) missiles, having retired the Polaris family of fleet ballistic

THE BALLISTIC MISSILE—A REVOLUTION IN WARFARE

In the middle of the 20th century, developments in rocket technology revolutionized warfare, transformed international politics, and changed forever our planetary civilization. One of the earliest milestones involved the combination of two powerful World War II–era weapon systems, the American nuclear bomb and the German V-2 ballistic missile. Cold war politics encouraged a hasty marriage of these emerging military technologies. The union ultimately produced an offspring called the intercontinental ballistic missile (ICBM)—a weapons system that most military experts regard as the single most influential weapon in all of history.

The ICBM and its technical cousin, the submarine-launched ballistic missile (SLBM), were the first weapons systems designed to attack a distant target by traveling into and through outer space. In the late 1950s, the first generation of such operational "space weapons" completely transformed the nature of strategic warfare and altered the practice of geopolitics.

For the United States, the development of the ICBM created a fundamental shift in national security policy. Before the ICBM, the major purpose of the American military establishment was to fight and win wars. With operational, nuclear weapon-armed ballistic missile forces, both the United States and the former Soviet Union possessed an unstoppable weapons system capable of delivering megatons of destruction to any point on the globe within about 30 or so minutes. Once fired, these weapons would fly to their targets with little or no chance of being stopped. To make matters worse, national leaders could not recall a destructive ICBM strike, after the missiles were launched. So from that moment on, the chief purpose of the U.S. military establishment became how to avoid strategic nuclear warfare. American and Russian leaders quickly realized that any major exchange of unstoppable nuclear-armed ballistic missiles would completely destroy both adversaries and leave Earth's biosphere in total devastation.

In the era of the ICBM, there are no winners, only losers—should adversaries resort to strategic nuclear conflict. For the first time in human history, national leaders controlled a weapons system that could end civilization in less than a few hours. The modern ballistic missile made deterrence of nuclear war the centerpiece of American national security policy—a policy often appropriately referred to as mutually assured destruction (MAD).

Some military analysts like to refer to the ICBM as an unusable weapon. For example, the United States and Russia designed their strategic nuclear forces to survive a surprise first strike and still deliver a devastating second strike. Consequently, no matter which side fired their nuclear-armed ICBMs first, the other side would still be capable of firing a totally destructive retaliatory salvo and everybody would die. In this perspective, the ICBM is quite analogous to a societal suicide weapon. One cold war military analyst summarized MAD as, "He who shoots first, dies second." Such gallows humor only serves to highlight the often-overlooked paradox. The most destructive weapons system in all of history is also its most unusable weapons system.

Until now, rational governments have tightly guarded the use of their nuclear-armed ICBM forces with elaborate control systems that restricted access to launch codes and nuclear weapon–arming signals. The arrival of the nuclear-armed ballistic missile is a milestone in human development. From a planetary culture perspective, this technical event brings about the end of childhood—that is, the end of social innocence. Will the family of nations mature into a global society capable of avoiding strategic nuclear warfare as a means of settling political, social, or ethnic differences, or will the human race perish as a technically advanced, but socially immature species?

Since the laws of physics that allowed the development of the nuclear-armed ICBM apply throughout the universe, it is interesting to speculate whether other technically advanced species in the Milky Way reached the same crossroads. Did such a species survive the "ICBM technology" test, or did it perish by its own hand?

An alternate technology pathway also appeared with the arrival of the operational ICBM. Cold-war politics encouraged the pioneering use of powerful ballistic missiles as space launch vehicles (SLVs). So, during the macabre superpower nuclear standoff that dominated geopolitics throughout most of the second half of the 20th century, the human race also acquired the very tools it needed to expand out into the solar system and mature as a planetary civilization. Will tribal barbarism finally give way to the rise of a mature, conflict-free global society? The ICBM is the social and technical catalyst capable of causing great harm or long-term good. Human beings have the choice to take the pathway that leads to the stars and not the one that leads to extinction.

Perhaps, as a species, the human race is now beginning to go down the right road. It is interesting to observe, for example, that the threat of instant nuclear Armageddon has helped restrain those nations with announced nuclear weapon capabilities (such as the United States, the former Soviet Union, the United Kingdom, France, and the People's Republic of China) from actually using such devastating weapons in resolving lower-scale regional conflicts. Because of the "no-winner" standoff between unstoppable offensive ballistic missile forces, political scientists assert that the ICBM has caused a revolution in warfare and international relations. Some military analysts even go so far as to suggest that the ICBM makes strategic warfare between rational, nuclear-armed nations impossible.

The operative term here is "rational." Today, as nuclear weapon and ballistic missile technologies spread throughout the world, the specter of a regional nuclear conflict (such as between India or Pakistan), nuclear blackmail (as threatened by the leadership of North Korea), or nuclear terrorism (at the hands of a technically competent, financially capable subnational political group) haunts the global community. The heat of long-standing, culturally or religiously rooted regional animosities could overcome the positive example of decades of self-imposed superpower restraint on the use of nuclear-armed ballistic missiles. Today's expanding regional missile threat is encouraging some American strategic planners to revisit ballistic missile defense technologies, including concepts requiring the deployment of antimissile weapon systems in outer space.

For more than four decades, the intercontinental ballistic missile has served as the backbone of the United States' nuclear deterrent forces. Throughout the cold war and up to the present day, deterring nuclear war has remained the top American defense priority. But this policy produces only a metastable political equilibrium that is based upon the concept of deterrence, which depends upon an inherent lose-lose outcome for its efficacy. What happens to the deterrence equation if an adversary who possesses nuclear-armed ballistic missiles does not care if they lose (for whatever unfathomable reason)?

Before the human race can enjoy the prosperity of a solar system civilization, governments everywhere must learn how to guarantee absolute control of the powerful ballistic missile weapons systems they now possess. Humankind's childhood has ended and people must face future political disagreements with unprecedented maturity. Otherwise, they will perish as a result of social immaturity and foolhardiness. The expanding global presence of nuclear-armed ballistic missiles offers no other alternatives.

missiles. The Trident II (D-5) FBM was first deployed in 1990 on the USS *Tennessee* (SSNB 734). The Trident II (D-5) FBM is a three-stage, solid propellant, inertially guided FBN with a range of more than 4,600 miles (7,400 km). The new Trident/Ohio class fleet ballistic missile submarines each carry 24 Trident IIs. These missiles can be launched while the submarine is underwater or on the surface. This type of ICBM is sometimes called a submarine-launched ballistic missile (SLBM).

Space Launch Vehicles

A space launch vehicle is an expendable or reusable rocket-propelled vehicle that provides sufficient thrust to place a spacecraft into orbit around earth or to send a payload on an interplanetary trajectory to another celestial body. Sometimes it is called a booster or space lift vehicle. This chapter explores the basic functions of a launch vehicle and provides examples of some historically significant or currently operational rockets used to place objects into outer space.

When rocket scientists want to place a spacecraft into orbit, the space launch vehicle they select has to successfully accomplish two important tasks: vertical ascent up through Earth's atmosphere and, once in space, horizontal (tangential) acceleration to orbital speed. First, the rocket vehicle must provide enough thrust to lift itself and its payload while ascending vertically up through the atmosphere. Second, once the rocket vehicle (or its final stage) reaches an appropriate altitude above Earth's surface (nominally at least 125 miles [200 km]), the rocket vehicle must pitch over (tilt) and provide the spacecraft (payload) a sideways nudge that provides a sufficient speed to keep it in orbit—that is, quite literally "falling around Earth." Scientists call this special velocity the orbital velocity and the spacecraft's closed path around Earth an orbit. If the rocket engines give the payload even more speed, the payload might reach escape velocity and depart on a trajectory into interplanetary space.

Aerospace engineers launch a large rocket vertically so the vehicle can travel the minimum distance necessary through the denser portions of the lower atmosphere at progressively increasing speeds that are still much slower that orbital velocity. This minimizes the adverse consequences of excessive aerodynamic force and frictional heating on the launch vehicle's structure and avoids damage or breakup. Any protective shrouds or fairing around the payload are usually jettisoned as the rocket departs the sensible atmosphere. (A fairing is a structural component of a rocket designed to reduce air resistance by smoothing out nonstreamlined objects or sections.) Then, as the rocket reaches outer space, the vehicle pitches over and provides the final velocity increment necessary for the payload to achieve orbit around Earth.

✧ The Origin of Space Launch Vehicles

While Konstantin Tsiolkovsky, Robert Goddard, and Hermann Oberth were the first people to envision independently the important role of the rocket in

WERNHER VON BRAUN
(1912–1977)

The German–American rocket engineer Wernher von Braun turned the impossible dream of interplanetary space travel into a reality. He started by developing the first modern ballistic missile, the liquid–fueled V–2 rocket, for the German army. He then assisted the United States by developing a family of ballistic missiles for their army and later a family of powerful space launch vehicles for NASA. One of his giant Saturn V rockets sent the first human explorers to the Moon's surface in July 1969, as part of the NASA's Apollo Project.

During the period between the 1930s and the 1970s, von Braun was one of the most important rocket developers and champions of space exploration. As a youth he became enamored with the possibilities of space exploration by reading the science fiction of Jules Verne. However, it was the science fact writings of Hermann Oberth, especially the 1923 classic *The Rocket into Interplanetary Space,* which prompted young von Braun to master calculus and trigonometry so he could understand the physics of rocketry.

From his teenage years, von Braun had held a keen interest in spaceflight, becoming involved in the German rocket society, Verein für Raumschiffahrt (VfR), as early as 1929. As a means of

furthering his desire to build large and capable rockets, in 1932 he went to work for the German army to develop ballistic missiles. While engaged in this work, he received a Ph.D. in engineering in July 1934.

Von Braun was the technical leader of what historians call the "rocket team," which developed the V–2 ballistic missile for the German army during World War II. They worked at a secret laboratory near Peenemünde on the Baltic coast. The V–2 rocket was the first modern liquid-propellant ballistic missile and the technical ancestor of the rockets used in space exploration programs in the United States and the former Soviet Union. Initially test flown as the A-4 rocket in October 1942, the V–2 was used as a military weapon against targets in Europe beginning in September 1944. By early 1945, it was obvious to von Braun that Nazi Germany would not achieve victory against the Allies, so he began planning for rocket research in the postwar era.

Before the Soviet army captured the V–2 rocket complex at Peenemünde, von Braun arranged for the escape and collective surrender of 500 of his top rocket scientists to American forces. For the next 15 years, von Braun worked with the United space travel, two rocket engineers, Wernher von Braun and Sergei Korolev, helped make the dream of rocket-propelled spaceflight into a reality.

Space launch vehicles come in a wide range of sizes and capabilities. The figure on page 76 shows the major U.S. space launch vehicles that supported space exploration in the 20th century.

The person who started the space age, on October 4, 1957, was the Russian rocket engineer, Sergei Korolev. He received permission to convert a powerful Soviet intercontinental ballistic missile (ICBM), called the R-7, into a space launch vehicle, and then placed *Sputnik 1*, the world's first artificial satellite, into orbit.

States Army in the development of ballistic missiles. As part of a U.S. military operation called Project Paperclip, he and his rocket team were scooped up from the ruins of a defeated Germany and relocated to Fort Bliss, Texas. Soon, at the White Sands Missile Range, they resumed launching V–2 rockets that had been captured at the end of the war. They also worked on the design of new ballistic missiles. In 1950, von Braun's team moved to the Redstone Arsenal near Huntsville, Alabama, where they developed the army's Redstone and Jupiter ballistic missiles. Teaming with the American entertainment genius Walt Disney, von Braun also served as one of the most prominent and well-recognized advocates for space exploration in the United States during the 1950s.

In 1960, von Braun's rocket development center was transferred from the U.S. Army to the newly established NASA. Soon, he received a mandate from NASA to build powerful "Moon-mission" rockets. Accordingly, von Braun became director of NASA's Marshall Space Flight Center and the chief architect of the Saturn V launch vehicle—the colossal booster that successfully propelled American astronauts to the Moon.

In 1970, NASA leadership asked von Braun to move to Washington, D.C., to head up the strategic planning effort for the agency. He left his home in Huntsville, Alabama, but in less than two years he decided to retire from NASA and to go to work for Fairchild Industries of Germantown, Maryland. He died in Alexandria, Virginia, on June 16, 1977.

Wernher von Braun in front of the Saturn 1B launch vehicle on the pad at Cape Canaveral, Florida (circa January 1968) *(NASA/Marshall Space Flight Center)*

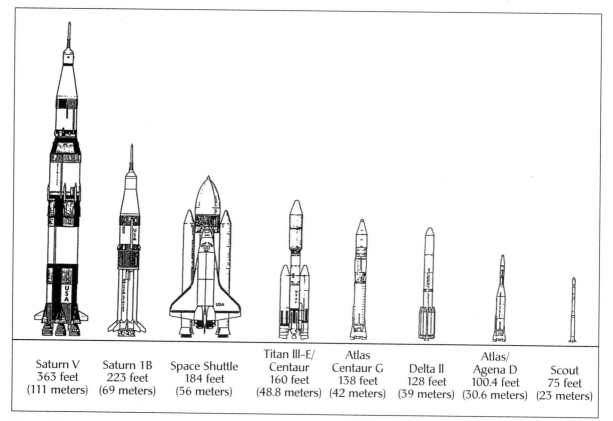

| Saturn V 363 feet (111 meters) | Saturn 1B 223 feet (69 meters) | Space Shuttle 184 feet (56 meters) | Titan III-E/ Centaur 160 feet (48.8 meters) | Atlas Centaur G 138 feet (42 meters) | Delta II 128 feet (39 meters) | Atlas/ Agena D 100.4 feet (30.6 meters) | Scout 75 feet (23 meters) |

Major U.S. launch vehicles that supported space exploration in the 20th century *(NASA)*

SERGEI KOROLEV
(1907–1966)

The Russian (Ukraine-born) rocket engineer Sergei Korolev was the driving technical force behind the initial intercontinental ballistic missile (ICBM) program and the early outer-space exploration projects of the former Soviet Union. In 1954, he started work on the first Soviet ICBM, the R-7. This powerful rocket system was capable of carrying a massive payload across continental distances. As part of cold-war politics Soviet premier Nikita Khrushchev (1894–1971) allowed Korolev to use this military rocket to place the first artificial satellite (*Sputnik 1*) into orbit around Earth on October 4, 1957. This event is now generally regarded as the beginning of the space age.

Korolev was trained in aeronautical engineering at the Kiev Polytechnic Institute and, after receiving a secondary education, cofounded the Moscow rocketry organization GIRD (Gruppa Isutcheniya Reaktivnovo Dvisheniya, Group for

Investigation of Reactive Motion). Like the VfR (Verein für Raumschiffahrt, Society for Spaceship Travel) in Germany, and Robert H. Goddard in the United States, the Russian organizations were by the early 1930s testing liquid-fueled rockets of increasing size. In Russia, GIRD lasted only two years before the military, seeing the potential of rockets, replaced it with the RNII (Reaction Propulsion Scientific Research Institute). RNII developed a series of rocket-propelled missiles and gliders during the 1930s, culminating in Korolev's RP-318, Russia's first rocket-propelled aircraft. Before the aircraft could make a rocket-propelled flight, however, Korolev and other aerospace engineers were thrown into the Soviet prison system in 1937–38, during the peak of Joseph Stalin's political purges.

Korolev at first spent months in transit on the Trans-Siberian railway and on a prison vessel at Magadan. This was followed by a year in the Kolyma gold mines, the most dreaded part of the gulag. However, Stalin recognized the importance of aeronautical engineers in preparing for the impending war with Hitler, and retrieved Korolev and other technical personnel from incarceration. He reasoned that these prisoners could help the Red Army by developing new weapons. Consequently, a system of sharashkas (prison design bureaus) was set up to exploit the jailed talent. Korolev was saved by the intervention of senior aircraft designer Sergei Tupolev, himself a prisoner, who personally requested Korolev's services in the TsKB-39 sharashka.

Following World War II, Korolev was released from prison and appointed chief constructor for development of a long-range ballistic missile. On April 1, 1953, as Korolev was preparing for the first launch of the R-11 rocket, he received approval from the Council of Ministers for development of the world's first ICBM, the R-7. To concentrate on development of the R-7, Korolev's

This 1969 postage stamp from the former Soviet Union celebrates the contributions of the Russian (Ukraine-born) rocket engineer Sergei Korolev—the person who started the Space Age by launching *Sputnik 1*, the world's first artificial Earth satellite. [Historic note: By the modern civil calendar, Korolev was born on January 12, 1907, in Zhitomir, Ukraine—at the time part of czarist Russia. However, the stamp indicates Korolev's birth date as December 30, 1906—a date corresponding to an obsolete czarist-era calendar system.] *(Courtesy of the author)*

other projects were spun off to a new design bureau in Dnepropetrovsk headed by Korolev's assistant, Mikhail Kuzmich Yangel. This was the first of several design bureaus that would spin off from Korolev's work. It was Korolev's R-7 ICBM that launched *Sputnik 1* on October 4, 1957. This historic launch also served to galvanize American concern about the capability of the Soviet Union to attack the United States with nuclear weapons using ballistic missiles. During the early 1960s, Korolev campaigned to send a Soviet cosmonaut to the Moon.

Following the initial reconnaissance of the Moon by the *Luna 1, 2,* and *3* spacecraft, Korolev established three largely independent efforts aimed at achieving a Soviet lunar landing

(continues)

(continued)

before the Americans. The first objective, met by *Vostok* and *Voskhod* spacecraft, was to prove that human space flight was possible. The second objective was to develop lunar vehicles, which would soft-land on the Moon's surface to ensure that a cosmonaut would not sink in-to the dust accumulated by four billion years of meteorite impacts. The third objective, and the most difficult to achieve, was to develop a huge booster to send cosmonauts to the Moon. Beginning in 1962, his design bureau began work on the N-1 launch vehicle, a counterpart to the American Saturn V. This giant rocket was to be capable of launching a maximum of 110,000 pounds (50,000 kg) into low-Earth orbit. Although the project continued until 1971, without Korolev's leadership the N-1 never made a successful flight.

On January 14, 1966, he died during a botched routine surgery at a hospital in Moscow. He was only 58 years old. Some of Korolev's contributions to space technology include: the powerful, legendary R-7 rocket (1956), the first artificial satellite (1957), pioneering lunar spacecraft missions (1959), the first human spaceflight (1961), a spacecraft to Mars (1962), and the first space walk (1965). Even after his death, the Soviet government chose to hide Korolev's identity by publicly referring to him only as the "chief designer of carrier rockets and spacecraft." Despite this official anonymity, chief designer and academician Korolev now is properly recognized as the brilliant rocket engineer who ushered in the space age.

As a result of the superpower competition in space exploration during the cold war, many military ballistic missiles were pressed into service as space launch vehicles. Other launch vehicles, like NASA's Saturn V "Moon rocket," were developed from the start for sending payloads and astronauts into space.

✧ Atlas

Atlas is a family of versatile liquid-fuel rockets originally developed by General Bernard Schriever of the United States Air Force in the late 1950s as the first operational American intercontinental ballistic missile (ICBM). Evolved and improved Atlas launch vehicles have served many government and commercial space transportation needs.

The modern Atlas rocket vehicle fleet, as developed and marketed by the Lockheed Martin Company, included three basic families: the Atlas II (IIA and IIAS), the Atlas III (IIIA and IIIB), and the Atlas V (400 and 500 series). The Atlas II family was capable of lifting payloads ranging in mass from approximately 6,160 pounds (2,800 kg) to 8,140 pounds (3,700 kg) to a geostationary transfer orbit (GTO). The Atlas III family of launch vehicles can lift payloads of up to 9,900 pounds (4,500 kg) to a geostation-

The Atlas–Centaur launch vehicle, carrying NASA's *Surveyor 1* spacecraft, lifts off from Pad 36A at Cape Canaveral Air Force Station, Florida, on May 30, 1966. Developed as the first American ICBM, the Atlas I rocket was also pressed into service as a versatile space launch vehicle for numerous military and civilian space missions. The *Surveyor 1* mission scouted the lunar surface in preparation for subsequent Apollo Project Moon-landing missions by American astronauts. *(NASA)*

ary transfer orbit, while the more powerful Atlas V family can lift payloads of up to 19,030 pounds (8,650 kg) to it.

The family of Atlas II vehicles evolved from the Atlas I family military missiles and cold war–era space launch vehicles. The transformation was brought about by the introduction of higher-thrust engines and longer propellant tanks for both stages. The Atlas III family incorporated the pressure-stabilized design of the Atlas II vehicle but used a new single-stage Atlas engine built by NPO Energomash of Russia. The RD-180 is a throttleable rocket engine that uses liquid oxygen and kerosene propellants and provides approximately 855,000 pounds-force (3,800,000 N) of thrust at sea level.

The Atlas V uses the RD-180 main engine in a Common Core Booster first stage configuration with up to five strap-on solid propellant rockets and a Centaur upper stage. The first Atlas V vehicle successfully lifted off from Cape Canaveral Air Force Station on August 21, 2002. Later that year, an important chapter in propulsion system history closed when the last Atlas II vehicle was successfully launched from Cape Canaveral Air Force Station on December 4, 2002.

✧ Delta

The Delta is a versatile family of American two- and three-stage liquid-propellant, expendable launch vehicles that uses multiple strap-on booster rockets in several configurations. Since May 1960, the Delta program has more than 270 successful military, civil, and commercial launches. Over the years, the Delta family of rocket vehicles has also accomplished many important space technology firsts. These achievements include the first international satellite, *Telstar I* (launched in 1962); the first geosynchronous satellite, *Syncom II* (launched in 1963); and the first commercial communications satellite, *COMSAT I* (launched in 1965). Because of its outstanding record of launch accomplishments, the Delta rocket has earned the nickname "space workhorse vehicle."

The evolving family of Delta rockets began in 1959, when NASA awarded a contract to Douglas Aircraft Company (now part of the Boeing Company) to produce and integrate 12 space launch vehicles. The first Delta vehicle used components from the United States Air Force's Thor missile for its first stage and from the United States Navy's Vanguard Project rocket for its second stage. On May 13, 1960, the inaugural flight of the Delta I rocket from Cape Canaveral Air Force Station successfully placed the *Echo 1* communications (relay) satellite (actually, a large, inflatable sphere) into orbit. The original Delta booster was 55.8 feet (17 m) long and 7.9 feet (2.4 m) in diameter. It used one Rocketdyne MB-3 Block

II rocket engine (also called the LR-79-NA11 engine) that burned liquid oxygen (LOX) and kerosene to produce 150,000 pounds-force (667,000 N) of thrust at liftoff. Two small vernier engines created 990 pounds-force (4,400 N) of thrust each, using the same propellants as the main engine. Continued improvements resulted in an evolutionary series of vehicles with increasing payload capabilities.

The Delta II is a medium-lift capacity, expendable launch vehicle used by the United States Air Force to launch the many satellites of the Global Positioning System (GPS). Additionally, the Delta II rocket launches civil and commercial payloads into low Earth orbit (LEO), polar orbit, geosynchronous transfer orbit (GTO), and geostationary orbit (GEO). NASA has used the versatile Delta II vehicle to launch many important scientific missions, including the Mars Exploration Rovers, *Spirit* and *Opportunity*.

The Delta II launch vehicle stands a total height of approximately 128 feet (39 m). In the configuration often used by the U.S. Air Force, the payload fairing—the protective shroud covering the third stage and the satellite payload—is 9.5 feet (2.9 m) wide to accommodate the GPS satellite. A stretched-version fairing is also available for larger payloads; it is 9.8 feet (3 m) wide. Six of the nine solid-rocket motors that ring the first stage separate after one minute of flight. The remaining three solid-rocket motors ignite and then separate after burnout about one minute later.

In the late 1990s, the Boeing Company introduced the Delta III family of rockets to serve space transportation needs within the expanding commercial communications satellite market. With a payload capacity of 8,360 pounds-mass (3,800 kg) delivered to a geosynchronous transfer orbit, the Delta III essentially doubled the performance of the Delta II rocket. The first stage of the Delta III is powered by a Boeing RS-27A rocket engine assisted by two vernier rocket engines that help control roll during main engine burn. The vehicle's second stage uses a Pratt and Whitney RL 10B-2 engine that burns cryogenic propellants.

On November 20, 2002, a Delta IV rocket lifted off from Cape Canaveral Air Force Station, successfully delivering the *W5 Eutelsat* commercial communications satellite to a geosynchronous transfer orbit. This event was the inaugural launch of the latest member of the Delta launch vehicle family. Delta IV rockets combine new and mature launch vehicle technologies and can lift medium to heavy payloads into space. The Delta IV vehicle also represents part of the United States Air Force's evolved expendable launch vehicle program. The new rocket uses the new Boeing Rocketdyne-built RS-68 liquid hydrogen and liquid oxygen main engine, an engine capable of generating 650,000 pounds-force (2,891,000 N) of thrust. Assembled in five vehicle configurations, the new Delta IV rocket family is capable of delivering payloads that range between 12,860 pounds-mass (5,845 kg) and 28,890 pounds-mass (13,130 kg) to a geosynchronous transfer orbit.

✧ Saturn

The Saturn family of expendable launch vehicles developed by Wernher von Braun's rocket team at NASA's Marshall Space Flight Center to support the Apollo Project. The Saturn 1B was used initially to launch Apollo lunar spacecraft into Earth orbit to help the astronauts train for the manned flights to the Moon. The first launch of a Saturn 1B vehicle with an unmanned Apollo spacecraft took place in February 1966. A Saturn 1B vehicle launched the first crewed Apollo flight, *Apollo 7*, on October 11, 1968. After completion of the Apollo Project, the Saturn 1B was used to launch the three crews (three astronauts each crew) for the *Skylab* space station in 1973. Then, in 1975, a Saturn 1B launched the American astronaut crew for the Apollo-Soyuz Test Project, a joint American-Russian docking mission. With an Apollo spacecraft on top, the Saturn 1B vehicle was approximately 223 feet (69 m) tall. This expendable launch vehicle developed 1.6 million pounds-force (7.1 million N) of thrust at liftoff.

The *Apollo 15* astronauts David R. Scott, Alfred M. Worden, and James B. Irvin lifted off aboard this Apollo/Saturn V rocket vehicle from Complex 39A at NASA's Kennedy Space Center on July 26, 1971. The colossal rocket generated 33.4 million newtons of thrust as it started the astronauts on their successful lunar landing mission. *(NASA)*

The Saturn V rocket, America's most powerful staged rocket, successfully carried out the ambitious task of sending astronauts to the Moon during the Apollo Project. The first launch of the Saturn V vehicle, the unmanned *Apollo 4* mission, occurred on November 9, 1967. The first crewed flight of the Saturn V vehicle, the *Apollo 8* mission, was launched in December 1968. This historic mission was the first human flight to the Moon. The three astronauts aboard *Apollo 8* circled the Moon (but did not land) and then returned safely to Earth. On July 16, 1969, a Saturn V vehicle sent the *Apollo 11* spacecraft and its crew on the first lunar landing mission. The last crewed mission of the Saturn V vehicle occurred on December 7, 1972, when the *Apollo 17* mission lifted off on the final human expedition to the Moon in the 20th century. The Saturn V vehicle flew its last mission on May 14, 1973, when it lifted the unmanned *Skylab* space station into Earth orbit. *Skylab* later was occupied by three different astronaut crews for a total period of 171 days.

All three stages of the Saturn V vehicle used liquid oxygen (LO_2) as the oxidizer. The fuel for the first stage was kerosene, while the fuel for the upper two stages was liquid hydrogen (LH_2). The Saturn V vehicle, with the Apollo spacecraft and its small emergency escape rocket on top, stood 363 feet (111 m) tall and developed 7.76 million pounds-force (34.5 million N) of thrust at liftoff.

✧ Scout

NASA uses the four-stage, all-solid-propellant Scout expendable launch vehicle to place small payloads into Earth orbit and to send probes on suborbital trajectories. The first Scout (Solid Controlled Orbital Utility Test) vehicle was launched on July 1, 1960, from the Mark 1 Launcher at the NASA Goddard Space Flight Center's Wallops Flight Facility in Wallops Island, Virginia.

The Scout's first-stage motor is based on an earlier version of the U.S. Navy's Polaris missile motor; the second-stage solid-propellant motor was derived from the U.S. Army's Sergeant surface-to-surface missile; and the third- and fourth-stage motors were adapted by NASA from the U.S. Navy's Vanguard missile.

The standard Scout vehicle is a slender, all-solid-propellant, four-stage booster system, approximately 75 feet (23 m) in length with a launch mass of 47,300 pounds (21,500 kg). This expendable vehicle is capable of placing a 409 pound-mass (186 kg) payload into a 350-mile (560-km) orbit around Earth.

✧ Titan

The Titan family of U.S. Air Force expendable launch vehicles was started in 1955. The Titan I missile was the first American two-stage intercontinental ballistic missile (ICBM) and the first underground silo–based ICBM. The Titan I vehicle provided many structural and propulsion techniques that were later incorporated in the Titan II vehicle. Years later the Titan IV evolved from the Titan III family and is similar to the Titan 34D vehicle.

The Titan II was a liquid-propellant, two-stage, intercontinental ballistic missile that was guided to its target by an all-inertial guidance and control system. This missile, designated as LGM-25C, was equipped with a nuclear warhead and designed for deployment in hardened and dispersed underground silos. The U.S. Air Force built more than 140 Titan ICBMs; at one time they served as the foundation of America's nuclear deterrent force during the cold war. The Titan II vehicles also were used as space launch vehicles in NASA's Gemini Project in the mid-1960s. Deactivation of the Titan II ICBM force began in July 1982. The last missile was taken from its silo at Little Rock Air Force Base, Arkansas, on June 23, 1987. The deactivated Titan II missiles then were placed in storage at Norton Air Force Base, California. Some of these retired ICBMs have found new use as space launch vehicles.

The Titan II space launch vehicle was a modified Titan II ICBM that was designed to provide low- to medium-mass launch capability into polar low Earth orbit (LEO). The modified Titan II was capable of lifting about 4,180 pounds (1,900 kg) into polar low Earth orbit. With the addition of two strap-on solid rocket motors (i.e., the graphite epoxy motors) to the first stage, the payload capability to polar low Earth orbit was increased to 7,770 pounds (3,530 kg).

The versatile Titan III vehicles supported a variety of defense and civilian space launch needs. The Titan IIIA vehicle was developed to test the integrity of the inertially guided three-stage Titan IIIC liquid-propulsion system core vehicle. The Titan IIIB consists of the first two stages of the core Titan III vehicle (without the two strap-on solid rocket motors) with an Agena vehicle used as the third stage. The Titan IIIC consisted of the Titan IIIA core vehicle with two solid rocket motors (sometimes called "Stage 0") attached on opposite sides of the liquid-propellant core vehicle. This configuration was developed for space launches from Complex 40 at Cape Canaveral Air Force Station, Florida. The Titan IIID vehicle, launched from Vandenberg Air Force Base, California, was essentially a Titan IIIC configuration without the Transtage and was radio-guided during launch. The Titan IIIE configuration was developed for NASA and is launched from Complex 41 at Cape Canaveral Air Force Station. The Titan IIIE was basically a standard Titan IIID with a Centaur vehicle used as the third stage. Finally, the Titan 34D configuration uses a stretched core vehicle in conjunction with larger solid rocket motors to increase booster performance.

The Titan IV was the newest and largest member of this rocket family. The U.S. Air Force developed the expendable Titan IV vehicle to launch the nation's largest high-priority, high-value "shuttle-class" defense payloads. This "heavy-lift" vehicle was flexible in that it could be launched with one of several optional upper-stages (such as the Centaur and the Inertia Upper Stage) for greater and more varied spacelift capability. The Titan IV vehicle's first stage consisted of an LR87 liquid-propellant rocket that featured structurally independent tanks for its fuel (aerozine 50) and oxidizer (nitrogen tetroxide). This design minimized the hazard of the two propellants mixing if a leak developed in either tank. Additionally, these liquid propellants were stored at normal temperature and pressure, eliminating launch pad delays (as often encountered with the boil-off and refilling of cryogenic propellants) and giving the Titan IV vehicle the capability to meet critical defense program and planetary mission launch windows.

The Titan IV vehicle had a length of up to 201 feet (61.2 m) and could carry payloads of up to 38,610 pounds (17,550 kg) into a 90-mile (145-km) altitude orbit when launched from Cape Canaveral Air Force Station; or up to 30,690 pounds (13,950 kg) into a 100-mile (160-km) altitude polar orbit when launched from Vandenberg Air Force Base, California.

The Titan IVB rocket was a heavy-lift space launch vehicle designed to carry government payloads, such as the Defense Support Program surveillance satellite or National Reconnaissance Office (NRO) satellites into space from Cape Canaveral Air Force Station, Florida, or Vandenberg Air Force Base, California. The Titan IVB could place 47,675 pounds (21,670 kg) into low Earth orbit or more than 12,670 pounds (5,760 kg) into geosynchronous orbit. The first Titan IVB rocket was successfully flown from Cape Canaveral Air Force Station on February 23, 1997. A powerful Titan IV-Centaur configuration successfully sent NASA's *Cassini* spacecraft to Saturn from Cape Canaveral Air Force Station on October 15, 1997.

✧ U.S. Space Transportation System

The U.S. Space Transportation System (STS) is NASA's official name for the overall space shuttle program, including intergovernmental agency requirements and international and joint projects. The major components of the space shuttle system are the winged orbiter vehicle (often referred to as the space shuttle); the three space-shuttle main engines; the giant external tank, which feeds liquid hydrogen fuel and liquid oxygen (oxidizer) to the shuttle's three main engines; and the two solid rocket boosters (SRBs).

The orbiter is the only part of the space shuttle system that has a name in addition to a part number. The first orbiter built was the *Enterprise* (OV-101), which was designed for flight tests in the atmosphere rather than operations in space. It is now at the Smithsonian Museum at Dulles Airport outside Washington, D.C. Five operational orbiters were constructed (listed in order of completion): *Columbia* (OV-102), *Challenger* (OV-99), *Discovery* (OV-103), *Atlantis* (OV-104), and *Endeavour* (OV-105). The *Challenger* and its crew were lost in a launch accident on January 28, 1986. The *Columbia* and its crew were lost in a reentry accident on February 1, 2003.

Shuttles are launched from either Pad 39A or 39B at the Kennedy Space Center, Florida. Depending on the requirements of a particular mission, a space shuttle can carry about 49,900 pounds (22,680 kg) of payload into low Earth orbit (LEO). An assembled shuttle vehicle has a mass of about 4,500,000 pounds (2,040,000 kg) at liftoff.

The two solid rocket boosters (SRBs) are each 149 feet (45.4 m) high, 12.1 feet (3.7 m) in diameter, with a mass of about 1,298,000 pounds

Liftoff of the space shuttle *Discovery* and its five-person crew from pad 39-B at the Kennedy Space Center on September 29, 1988. This was the start of the successful STS-26 mission—the return-to-flight mission that followed the 1986 *Challenger* accident. *(NASA)*

(590,000 kg). Their solid propellant consists of a mixture of powdered aluminum (fuel), ammonium perchlorate (oxidizer), and a trace of iron oxide to control the burning rate. The solid mixture is held together with a polymer binder. Each booster produces a thrust of approximately 3.1 million pounds-force (13.8 million N) for the first few seconds after ignition. The thrust then gradually declines for the remainder of the two-minute burn, to avoid overstressing the flight vehicle. Together with the three main liquid-propellant engines on the orbiter, the shuttle vehicle produces a total thrust of 7.3 million pounds-force (32.5 million N) at liftoff.

Typically, the SRBs burn until the shuttle flight vehicle reaches an altitude of about 28 miles (45 km) and a speed of 3,090 miles per hour (4,970 km/h). Then they separate and fall back into the Atlantic Ocean to be retrieved, refurbished, and prepared for another flight. After the solid rocket boosters are jettisoned, the orbiter's three main engines, fed by the huge external tank, continue to burn and provide thrust for another six minutes before they too are shut down at MECO (main engine cutoff). At this point, the external tank is jettisoned and falls back to Earth, disintegrating in the atmosphere with any surviving pieces falling into remote ocean waters.

The huge external tank is 154 feet (47 m) long and 27.6 feet (8.4 m) in diameter. At launch, it has a total mass of about 1,672,550 pounds (760,250 kg). The two inner propellant tanks contain a maximum of 385,000 gallons (1,458,400) liters of liquid hydrogen (LH_2) and 143,400 gallons (542,650 liters) of liquid oxygen (LO_2), respectively. The external tank is the only major shuttle flight vehicle component that is expended on each launch. Following the loss of the *Columbia* in February 2003, the external tank underwent major design changes to minimize the generation of launch debris that could damage the orbiter vehicle.

The winged orbiter vehicle is both the heart and the brains of NASA's space shuttle. About the same size and mass as a commercial DC-9 jet aircraft, the orbiter contains the pressurized crew compartment (which can normally carry up to eight crew members), the huge cargo bay (which is 60 feet [18.3 m] long and 15 feet [4.57 m] in diameter), and the three main engines mounted on its aft end. The orbiter vehicle itself is 121 feet (37 m) long and 56 feet (17 m) high and has a wingspan of 79 feet (24 m). Since each of the operational vehicles varies slightly in construction, an orbiter generally has an empty mass that ranges from 167,200 pounds (76,000 kg) to 173,800 pounds (79,000 kg).

Each of the three main engines on an orbiter vehicle is capable of producing a thrust of 375,300 pounds-force (1,668,000 N) at sea level and 470,250 pounds-force (2,090,000 N) in the vacuum of space. These engines burn for approximately eight minutes during launch ascent and together consume about 64,000 gallons (242,250 L) of cryogenic propellants each minute, when all three operate at full power.

An orbiter vehicle also has two smaller orbital maneuvering system (OMS) engines that operate only in space. These engines burn nitrogen tetroxide as the oxidizer and monomethyl hydrazine as the fuel. These propellants are supplied from onboard tanks carried in the two pods at the upper rear portion of the vehicle. The OMS engines are used for major maneuvers in orbit and to slow the orbiter vehicle for reentry at the end of its mission in space. On most missions the orbiter enters an elliptical orbit, then coasts around Earth to the opposite side. The OMS engines then fire just long enough to stabilize and circularize the orbit. On some missions the OMS engines also are fired soon after the external tank separates, to place the orbiter vehicle at a desired altitude for the second OMS burn that then circularizes the orbit. Later OMS engine burns can raise or adjust the orbit to satisfy the needs of a particular mission. A shuttle flight can last from a few days to more than a week or two.

After deploying the payload spacecraft (some of which can have attached upper stages to take them to higher-altitude operational orbits, such as a geostationary orbit), operating the onboard scientific instruments, making scientific observations of Earth or the heavens, or performing other aerospace activities, the orbiter vehicle reenters Earth's atmosphere and lands. This landing usually occurs at either the Kennedy Space Center in Florida (primary site) or at Edwards Air Force Base in California—depending on weather conditions at the primary landing site. Unlike prior manned spacecraft, which followed a ballistic trajectory, the orbiter (now operating like an unpowered glider) has a cross-range capability of about 1,240 miles (2,000 km)—that is, it can move to the right or left off the straight line of its reentry path. The landing speed is between 210 miles per hour (340 km/h) and 225 miles per hour (365 km/h). After touchdown and rollout, the orbiter vehicle immediately is "safed" by a ground crew with special equipment. This safing operation is also the first step in preparing the orbiter for its next mission in space.

The orbiter's crew cabin has three levels. The uppermost is the flight deck, where the commander and pilot control the mission. The middeck is where the galley, toilet, sleep stations, and storage and experiment lockers are found. Also located in the middeck are the side hatch for passage to and from the orbiter vehicle before launch and after landing and the airlock hatch into the cargo bay and to outer space to support on-orbit extravehicular activities. Below the middeck floor is a utility area for air and water tanks.

The orbiter's large cargo bay is adaptable to numerous tasks. It can carry satellites, large space platforms such as the *Long-Duration Exposure Facility (LDEF)*, and even an entire scientific laboratory, such as the European Space Agency's *Spacelab,* to and from low Earth orbit (LEO). It also serves as a workstation for astronauts to repair satellites, a foundation from which to erect space structures, and a place to store and hold spacecraft that have been retrieved from orbit for return to Earth.

Mounted on the port (left) side of the orbiter's cargo bay behind the crew quarters is the remote manipulator system (RMS), which was developed and funded by the Canadian government. The RMS is a robot arm and hand with three joints similar to those found in a human being's shoulder, elbow, and wrist. There are two television cameras mounted on the RMS near the "elbow" and "wrist." These cameras provide visual information for the astronauts who are operating the RMS from the aft station on the orbiter's flight deck. The RMS is about 49 feet (15 m) in length and can move anything, from astronauts to satellites, to and from the cargo bay as well as to different points in nearby outer space.

NASA has safety-upgraded the remaining Orbiter vehicles—namely *Discovery, Atlantis* and *Endeavour.* Following a hiatus of more than two years, NASA personnel successfully launched the space shuttle *Discovery* on July 26, 2005, from the Kennedy Space Center in Florida. After docking with the *International Space Station,* the *Discovery* returned to Earth and landed at Edwards AFB, in California, on August 9. However, the STS-114 mission was overshadowed by the problem of pieces of insulating foam peeling off the external tank during liftoff. Fortunately, as detailed on-orbit inspection by astronauts indicated, the *Discovery* had not received any major structural damage and was able to safely return to Earth. But the STS-114 mission tank insulation incident prompted NASA officials to ground again the shuttle fleet, while engineers searched for a better solution to this serious safety problem.

Hampered by continued safety problems with the shuttle fleet, NASA personnel are now pursuing the development of two new launch vehicles, and they plan to retire the shuttle as soon as possible after completion of the *International Space Station.* On September 19, 2005, the NASA administrator, Michael Griffin, introduced a new spacecraft designed to carry four astronauts to and from the Moon, to support up to six crew members on future missions to Mars, and to deliver crew and supplies to the *International Space Station.* The launch system that will get the crew off the ground builds on powerful, reliable shuttle propulsion elements. In about 2010, astronauts will launch on a rocket made up of a single shuttle solid rocket booster, with a second stage powered by a shuttle main engine. This launch system will be significantly safer than the shuttle because of an escape rocket on top of the crew capsule that can quickly blast the astronauts away if problems develop during launch. Furthermore, because the crew capsule sits on top of the rocket, there is little chance the capsule will experience

Around 2010, astronauts will ascend into space aboard this rocket for a rendezvous mission with the *International Space Station*. NASA's latest crew-carrying launch vehicle builds upon shuttle propulsion system technology and consists of a single shuttle solid rocket booster along with a second stage powered by a (liquid propellant) shuttle main engine. *(Artist's concept courtesy of NASA/ John Frassanito and Associates)*

any significant damage from launch vehicle debris in the event of an explosion.

The second new vehicle is a heavy-lift system that uses a pair of longer solid rocket boosters and five shuttle main engines to put up to 275,000 pounds (125,000 kg) into low Earth orbit. NASA plans to use this versatile heavy-lift system to place the components needed to go to the Moon and Mars into orbit around Earth.

✧ Soyuz

The Soyuz launch vehicle has served as the "workhorse" Soviet (and later Russian) launch vehicle. This vehicle evolved from the original R-7 ICBM developed by Sergei Korolev and his OKB-1 design bureau. The rocket was first used in 1963. With its two cryogenic stages and four cryogenic strap-on engines, this vehicle is capable of placing up to 15,180 pounds (6,900 kg) into low Earth orbit (LEO). At present, it is the most frequently flown launch vehicle in the world. Since 1964, the Soyuz rocket has been used to launch every Russian human crew space mission.

✧ Proton

The Proton is a Russian liquid-propellant expendable launch vehicle. The four-stage configuration of the rocket vehicle often is used for interplanetary missions. This vehicle consists of three hypergolic stages and one cryogenic stage and is capable of placing 10,540 pounds (4,790 kg) into a geostationary transfer orbit (GTO)—that is, into an elliptical orbit with an apogee of 22,000 miles (35,400 km) and a perigee of 450 miles (725 km). The Proton D-1 vehicle consists of three hypergolic stages

A Russian Proton rocket lifts off from the Baikonour Cosmodrome in Kazahkstan on November 20, 1998. This launch successfully placed the *Zarya* module—the first component of the *International Space Station*—into low Earth orbit. *(NASA)*

and can place up to 46,100 pounds (20,950 kg) into low Earth orbit (LEO). The three-stage Proton vehicle configuration was used to place *Salyut* and *Mir* space station modules into orbit. A Proton rocket vehicle also placed the *Zarya* module—the first component of the *International Space Station*—into orbit in November 1998.

An Ariane 5 expendable launch vehicle lifts off from ELA-3 at the Guiana Space Center in Kourou, French Guiana on December 10, 1999. This first operational flight of the Ariane 5 launch vehicle successfully carried the European Space Agency's X-Ray Multi-Mirror Mission observatory into orbit. *(ESA/CNES/Arianespace-Service Optique CSG; Copyright © ESA; used with permission)*

✧ Ariane

The Ariane family of launch vehicles evolved from a European desire to achieve and maintain an independent access to space. Early manifestation of the efforts that ultimately resulted in the creation of Arianespace (the international company that now markets the Ariane launch vehicles) included France's Diamant launch vehicle program (with operations in the Sahara Desert at Hammaguir, Algeria) and the Europa launch vehicle program, which operated in the Woomera Range in Australia before moving to Kourou, French Guiana, in 1970.

These early efforts eventually yielded the first Ariane flight on December 24, 1979. That mission, called L01, was followed by 10 more Ariane-1 flights over the next six years. The initial Ariane vehicle family (Ariane-1 through Ariane-4) centered on a three-stage launch vehicle configuration with evolving capabilities. The Ariane-1 vehicle gave way in 1984 to the more powerful Ariane-2 and Ariane-3 vehicle configurations. These configurations, in turn, were replaced in June 1988 with the successful launch of the Ariane-4 vehicle—a launcher that has been called Europe's "space workhorse." Ariane-4 vehicles were designed to orbit satellites with a total mass value of up to 10,340 pounds (4,700 kg). The various launch vehicle versions differ according to the number and type of strap-on boosters and the size of the fairings.

On October 30, 1997, a new, more powerful launcher, called Ariane-5, joined the Ariane family when its second qualification flight (called V502 or Ariane 502) successfully took place at the Guiana Space Center in Kourou. Twenty-seven minutes into this flight, the *Maqsat-H* and *Maqsat-B* platforms, carrying instruments to analyze the new launch vehicle's performance, and the *TeamSat* technology satellite were injected into orbit.

The Ariane-5 launch vehicle is an advanced "two-stage" system, consisting of a powerful liquid hydrogen/liquid oxygen-fueled main engine (called the Vulcain) and two strap-on solid propellant rockets. It is capable of placing a satellite payload of 12,980 pounds (5,900 kg) (dual satellite launch) or 14,960 pounds (6,800 kg) (single satellite launch) into geostationary transfer orbit (GTO), and approximately 44,000 pounds (20,000 kg) into low Earth orbit (LEO).

With the retirement of the Ariane-4 system, the continuously improving Ariane-5 propulsion system now serves as the main launch vehicle for the European Community. The Ariane-5 launch vehicle can place a wide variety of payloads into LEO, into geostationary Earth orbit (GEO), and on interplanetary trajectories.

Launch Sites

A launch site is the specific place from which aerospace industry personnel send a rocket into outer space. This chapter describes the general characteristics and features of a launch site, as well as some of the world's major launch complexes—including Cape Canaveral, Vandenberg Air Force Base, the Baikonur Cosmodrome, and the Guiana Space Center.

Aerospace engineers define the launch site as the geographic area from which they can conduct a test or operational launch of either a reusable aerospace vehicle or an expendable rocket. For safety reasons, rocket scientists select a remote location surrounded by large amounts of undeveloped land or bodies of water, so the rockets can fly without passing over inhabited areas. They also prefer a site with pleasant, mild weather conditions. Rockets departing from a site near the equator in an easterly 90° launch azimuth receive the maximum natural velocity boost from Earth's west-to-east rotation. (Launch azimuth is the horizontal direction or bearing of a body.) Therefore, launch sites located near Earth's equator—such as Guiana Space Center in Kourou, French Guiana, and Cape Canaveral, Florida—are especially valuable. Good transportation (by land, sea, and air) to the launch site helps avoid excessive shipping costs and prevents delays in the delivery of rocket vehicles, their payloads, propellants, support equipment, and other necessary supplies.

Rocket scientists and engineers use the term *launch complex* to describe the complete collection of launch sites (pads), control center, support facilities, and aerospace ground equipment used to launch expendable rockets and aerospace vehicles from a given geographic location. A launch complex, like NASA's Kennedy Space Center, can also include a landing strip to allow the return of an aerospace vehicle from its space mission. In this case, the launch complex serves as both a doorway to outer space from the surface of the planet and a port of entry from space to the planet's surface. Aerospace engineers call this facility a spaceport. In the future, as permanent human bases are established on the Moon and Mars, such bases will also have spaceports, which provide access to and from space and accommodate interplanetary travelers.

✧ Launch Site Functions

Here on Earth, aerospace personnel perform many different operations at a launch complex. They receive and assemble expendable launch vehicles, integrate payloads with their rocket vehicles, test and check out the combined payload/flight vehicle configuration, fuel the launch vehicle at the pad, coordinate weather, range safety, and tracking activities, and conduct countdown and launch operations. For reusable aerospace vehicles, like NASA's space shuttle, the complex also supports landing operations and vehicle refurbishment.

When aerospace engineers launch a rocket, they do so by performing a detailed process called the countdown. This is a step-by-step, carefully scheduled set of procedures that ultimately leads to the ignition of the rocket's engines. During the countdown of a modern rocket, aerospace personnel bring the flight vehicle to the launch site and load it with payload and propellants. Using launch-center computers that communicate with sensors onboard the rocket, they monitor all of the important systems on the launch vehicle and its payload. Aerospace personnel need to monitor both the rocket vehicle and its payload during the countdown, because it would be foolish to put a malfunctioning (or broken) object into space.

Launch personnel must also carefully monitor the weather and wait for the proper launch window. The launch window is that precise time interval during which aerospace personnel can launch a rocket so that the payload can reach the intended orbital destination. The launch crew must fire the rocket during this window, or else the mission will fail. If the launch crew gets rocket vehicles that use cryogenic propellants ready too soon, these propellants could boil off and the vehicle would need to be refueled. If they get a rocket vehicle ready too late, they miss the launch window—causing costly delays and the need to recycle the countdown—repeating much (if not all) of the time-consuming procedures.

Each countdown takes place in accordance with a specific predesignated schedule and has a detailed checklist of required events, actions, and conditions. Time is marked off (or counted down) in a reverse manner with zero representing the go or launch time. A number of factors can influence the countdown process, including weather conditions, unanticipated equipment glitches or malfunctions, unauthorized people entering the launch site or its companion safety zone, and the available launch window, which is often very mission dependent.

Weather at the launch site plays a major, but uncontrollable, role in the countdown. Unfavorable environmental conditions, such as high winds, thunderstorms, and the threat of lightning strikes, will delay a launch.

The presence of unauthorized persons who suddenly enter the safety "keep-out" (or exclusion) area at or proximate to the launch site can also interfere with a countdown.

During launch operations in the American aerospace program, personnel use "T-time" to reference specific times (plus or minus) with respect to the zero time (launch time) that occurs at the end of the countdown. Early in a countdown, for example, the rocket vehicle could arrive at the launch pad at "T minus 30 days"—aerospace industry jargon meaning 30 days before the scheduled launch day. As the countdown process gets closer to the actual launch time, support personnel will often hear important announcements like: "T minus 20 seconds and counting." This means that the countdown is going smoothly and that it is precisely 20 seconds from rocket engine ignition and liftoff.

If a problem develops during the countdown, the launch director will make an announcement like: "T minus 40 seconds and holding." This means that at precisely 40 seconds before the planned liftoff, launch personnel have suspended the countdown process. Aerospace workers use the term *hold* to identify a halt in the normal sequence of events that take place during a countdown. After launch personnel have resolved the problem (a minor problem is called a glitch), they usually pick up (or resume) the countdown from the point where they left off.

Once in a while, solving one problem during a countdown hold creates undesirable conditions that could lead to another problem—for example, cryogenic propellant boiloff. Under such circumstances, the launch director usually decides to backtrack in the countdown process just to make sure that no potential problems can adversely impact the launch. In such cases, he makes a decision to recycle the countdown from perhaps "T minus 40 seconds" to "T minus 80 seconds," or whatever minus T-time is appropriate. Following these directions, all launch personnel go back to their detailed list of instructions and resume the countdown by performing the activities specified for T minus 80 seconds—and proceed from that point.

Within the current American aerospace program, "planned holds" are customarily included in the countdown sequence. This occurs so the launch-support computers can run thousands of automatic checks on the rocket vehicle and its payload moments before engine ignition. Other times, an "unscheduled hold" is necessary because a problem suddenly appears. For example, launch site weather conditions might quickly go from good to marginal. The launch director would then declare a "weather hold" in order to allow the launch team personnel some time to fully examine the deteriorating environmental conditions before continuing with the countdown. When the launch director releases the hold, the countdown clock again begins to tick down to zero.

At zero time (that is, T-time equal to zero), the launch director sends the ignition signal to the rocket's engines, and they roar to life. As the vehicle rises from the pad, the countdown clock enters the positive (after liftoff) portion of T-time. For example, at approximately T plus two minutes, the space shuttle jettisons its expended solid rocket boosters. Once a space launch vehicle has done its job and its payload heads into orbit or on some desired interplanetary trajectory, flight control personnel begin using another convenient reference time called mission elapsed time (MET). This is the time that has gone by since the start of a space mission—usually taken as the moment of rocket vehicle liftoff.

The ride into space, from rocket vehicle liftoff to payload insertion into low Earth orbit, typically takes about 10 to 15 minutes—depending upon the rocket vehicle, launch site, and desired location for the payload in low Earth orbit. Then, if needed, an upper-stage rocket propulsion system (discussed in chapter 7) will send the payload to a higher altitude orbit around Earth or place the payload on an appropriate trajectory through interplanetary space.

✦ Scrubs, Flubs, and Range Safety

If a problem that defies immediate resolution occurs during the countdown, the launch director will postpone (or scrub) the launch for that particular day. Depending on the severity of the problem and the complexity of the launch vehicle, another countdown could start within a day, or a week or more might elapse before the launch team attempts to send the particular rocket vehicle into space. No prudent launch director will take an unnecessary risk with a very expensive space launch vehicle and its equally costly payload. Besides, when fully fueled and sitting on the pad, a large chemical rocket is a very temperamental and potentially dangerous object.

Sometimes, despite very attentive countdown procedures, a rocket misbehaves when a critical piece of equipment suddenly fails during an attempted launch. Aerospace personnel then have to abort (that is, cut short or cancel) the mission. Launch vehicle aborts can occur on the pad or in flight. In one type of on-pad abort, the engines of a fully fueled rocket fail to ignite. A "hang-fire" rocket is a very dangerous device that might explode suddenly without warning, so the launch team must exercise extreme caution in backing out of the countdown and securing the vehicle.

Each launch site has well-established safety guidelines for handling such on-pad hang-fires. If the rocket vehicle is carrying a human crew, their emergency evacuation is the first and most important activity. Other priority activities involve the securing of all electrical systems and circuits and carefully unloading any liquid propellants. To the greatest extent possible, the launch team personnel conduct these reverse countdown (or backout) procedures from a safe, remote location—assisted by automated equipment and teleoperated devices.

Sometimes a rocket vehicle's engines ignite, but then do not develop sufficient thrust. The vehicle struggles to get a few meters or so into the air, lingers suspended over the pad, and then settles on the ground, disappearing in a huge explosion. This is called an on-pad (or a near-pad) explosive abort. The violent explosion showers the immediate launch site with burning chemicals, shrapnel, and large chunks of debris. At a properly designed launch site, the personnel responsible for supervising the final stages of the countdown are safe in a distant control center, despite the explosive abort of a rocket.

Once a "live" rocket clears the pad, any serious malfunction creates an in-flight abort of the vehicle. Every operational rocket range has a command-destruct officer whose primary duty is to closely monitor the flight trajectory of an ascending rocket. If that rocket begins to veer off its planned course for whatever reason, this person prepares to send the rocket an encoded command-destruct signal. Once the trajectory of an errant rocket touches the destruct line (an imaginary boundary line clearly defined for each launch), the self-destruct command is sent to the misbehaving vehicle. This command destruct signal is an encrypted radio frequency signal, which activates the special self-destruction explosive system carried by every missile or rocket flying from an established range.

The use of a command-destruct system is an integral part of the range safety process. The shape and extent (that is, footprint) of the destruct line varies with the type of rocket being flown and the location of the launch site. Command destruct of an erratic rocket prevents it from endangering people and property outside the boundaries of the rocket range. Safety experts also use the destruct line to protect personnel and support facilities at the launch complex.

As a launch vehicle rises toward outer space, sometimes its second- or third-stage rocket malfunctions. In this case, there is usually no danger to people or property in the immediate vicinity of the launch site. Generally, a range's launch azimuth restrictions prevent downrange in-flight aborts from showering debris on inhabited regions. The launch azimuth is the initial compass heading of a powered rocket vehicle at launch. For example, range safety requirements established by Cape Canaveral Air Force Station require that all rockets launched from this site (including the Kennedy Space Center) fly on trajectories with a launch azimuth between 35 degrees and 120 degrees. This means that a rocket can leave the range by flying only in a northeasterly, easterly, or slightly southeasterly direction.

Flying due east (that is, on a 90-degree launch azimuth) is actually quite favorable, since the rocket vehicle picks up the full "free" velocity increment provided by Earth's natural west-to-east spin.

A range's launch azimuth restrictions also guarantee that debris impacts due to the normal jettisoning of expended booster stages, fairings, and other similar hardware stay within planned and acceptable hazard levels. A modern rocket range usually extends for thousands of kilometers away from the launch site. The idea is to minimize any impact hazard from normal launch vehicle debris as intentionally discarded equipment falls back to Earth. Planned impact zones in uninhabited, broad-ocean areas have proven quite suitable. Downrange tracking sta-

An aborted Juno II rocket, seconds before total explosion at Cape Canaveral Air Force Station on July 16, 1959 (United States Air Force)

tions (and, when appropriate, tracking ships) help aerospace personnel monitor the entire ascent trajectory of a modern launch vehicle. Should an upper stage fail in some unusual way, creating danger to people half a world away, personnel at a downrange station can also send an encoded command-destruct signal to the remaining propulsive portions of the flight vehicle.

✧ Major Launch Sites

Major launch sites in the United States include the Eastern Range at Cape Canaveral Air Force Station/Kennedy Space Center on the central east coast of Florida, the Western Range at Vandenberg Air Force Base on the west central coast of California, and NASA's Wallops Flight Facility on the east coast of Virginia. Major Russian launch sites include Plesetsk, Kapustin Yar, and the Baikonur Cosmodrome (also called Tyuratam), which is located in Kazakhstan. The European Space Agency has a major launch site on the northeast coast of South America in Kourou, French Guiana.

CAPE CANAVERAL AIR FORCE STATION

Cape Canaveral Air Force Station (CCAFS), on Florida's east-central coast along the Atlantic Ocean, is the world's most famous rocket range. It is the region from which the United States Air Force and NASA have launched more than 3,000 rockets since 1950. It is the first station in the 10,000-mile- (16,000-km-) long Eastern Range. Together with Patrick Air Force Base (some 20 miles [32 km] to the south), Cape Canaveral Air Force Station forms a complex that is the center for Department of Defense launch operations on the east coast of the United States. The adjacent NASA Kennedy Space Center serves as the spaceport for the fleet of space shuttle vehicles. CCAFS covers a 25-square-mile [65 km^2] area. Much of the land is now inhabited by large populations of animals, including deer, alligators, and wild boar. The nerve center for CCAFS and the entire Eastern Range is the range control center, from which all launches, as well as the status of range resources, are monitored. The range safety function also is performed in the range control center.

Launch sites include space launch complexes 40 and 41 in the Integrated Transfer Launch Facility (ITLF), where preparation and launch of powerful Titan rocket vehicles took place. (The final Titan rocket launch from Cape Canaveral occurred in 2005.) Delta rocket vehicles are processed and launched from space launch complexes 17A and 17B, while Atlas vehicles are launched from complexes 36A and 36B. Space launch complex 20 is used for suborbital launches, and meteorological sounding rockets are launched from the Meteorological Rocket Launch Facility.

Looking north at the rocket launch complexes of Cape Canaveral Air Force Station in the early 1970s. During the ballistic missile race of the cold war, this unique collection of launch facilities, situated along the shore of the Atlantic Ocean in central Florida, became known as "Missile Row." The Titan rocket launch complex (called Complex 40) appears in the distance. *(NASA and the United States Air Force)*

KENNEDY SPACE CENTER

NASA's John F. Kennedy Space Center (KSC) is immediately north and west of Cape Canaveral Air Force Station. The center is about 34 miles (55 km) long and varies in width from five miles (8 km) to 10 miles (16 km). The total land and water occupied by the installation is 140,393 acres (56,817 ha). Of this total area, 84,031 acres (34,007 ha) are NASA-owned. The remainder of the area is owned by the state of Florida. This large area, with adjoining bodies of water, provides the buffer space necessary to protect the nearby civilian communities during space vehicle launches. Agreements have been made with U.S. Department of the Interior supporting the use of the nonoperational (buffer) area as a wildlife refuge and national seashore. The complex honors John Fitzgerald Kennedy (1917–1963), who, as 35th president of the United States, committed the nation to the Apollo Project.

KSC was established in the early 1960s to serve as the launch site for the Apollo-Saturn V lunar-landing missions. After the Apollo Project ended in 1972, Launch Complex 39 was used to support both the *Skylab* program (the early nonpermanent U.S. space station) and then the Apollo-Soyuz Test Project (an international rendezvous and docking demonstration that involved spacecraft from the United States and the former Soviet Union).

The Kennedy Space Center now serves as the primary center within NASA for the test, checkout, and launch of space vehicles. The center's responsibility includes the launching of crewed (space shuttle) vehicles at launch complex 39 and NASA expendable launch vehicles (such as the Delta II rocket) at both nearby Cape Canaveral Air Force Station and at Vandenberg Air Force Base in California. The assembly, checkout, and launch of the space shuttle vehicles and their payloads take place at the center.

Weather conditions permitting, the orbiter vehicle lands at the Kennedy Space Center (after an orbital mission) and undergoes "turnaround," or processing between flights. The Vehicle Assembly Building (VAB) and the two shuttle launch pads at Launch Complex 39 may be the best-known structures at KSC, but other facilities also play critical roles in prelaunch processing of payloads and elements of the space shuttle system. Some buildings, such as the 525-foot- (160-m-) tall VAB, were originally designed for the Apollo Project in the 1960s and then altered to accommodate the space shuttle. Other facilities, such as the Orbiter Processing Facility (OPF) high bays, were designed and built exclusively for the space shuttle program.

Launch complex 39A at NASA's Kennedy Space Center (KSC), Florida (November 1982) *(NASA)*

Cape Canaveral—Pathway to the Stars

The long procession of rocket launches from Cape Canaveral mark a major milestone in the migration of conscious intelligence beyond the confines of our tiny planetary biosphere. With the first rocket launch from Cape Canaveral on July 24, 1950, the destiny of the human race changed forever. Consider the fact that on Earth, the last such major evolutionary unfolding occurred about 350 million years ago, when prehistoric fish, called crossopterygians, first left the ancient seas and crawled upon the land. Scientists regard these early "explorers" as the ancestors of all terrestrial animals with backbones and four limbs. Similarly, future galactic historians will note how life emerged out of Earth's ancient oceans, paused briefly on the land, then boldly ventured forth from Cape Canaveral to the stars.

The "Cape" is the most well-recognized space launch facility on the planet. It is generally considered the geographic area on the east central coast of Florida that contains Cape Canaveral Air Force Station, NASA's Kennedy Space Center, the Merritt Island National Wildlife Refuge, Cape Canaveral National Seashore, and Port Canaveral. Virtually uninhabited in the early 1950s, the remote location allowed rocket scientists and engineers to inspect, fuel, and launch rockets without endangering nearby communities. Test rockets fired from the Cape flew over open water instead of over populated land areas. The region's sunny, subtropical climate permitted year-round operations. And, as a bonus, this location provided a natural ("free") velocity boost to rockets launched eastward due to the west-to-east rotation of Earth.

With the birth of the space age on October 4, 1957, the military rocket testing equation became far more complicated. From a secret missile test facility in central Asia (now called the Baikonur Cosmodrome in the independent republic of Kazakhstan), the former Soviet Union used its most powerful military missile (called the R-7) to place the world's first artificial satellite, *Sputnik 1*, into orbit around Earth.

Rockets and missiles no longer belonged exclusively to the superpower arms race. Instead, as space launch vehicles, these rockets became highly visible instruments of global politics. During the cold war, superpower prestige became directly linked to technical achievements in space exploration. Caught up in the "space race," the Cape soon became a major technology showcase for Western democracy.

In the late evening (local time) of January 31, 1958, the ground near Launch Complex 26 shook as a hastily modified U.S. Army Juno rocket roared from its pad and successfully inserted the first American satellite, called *Explorer 1*, into orbit around Earth. This effort was the culmination of a joint project of the U.S. Army Ballistic Missile Agency in Huntsville, Alabama, and the Jet Propulsion Laboratory (JPL) in Pasadena, California. Wernher von Braun supervised the rocket team, while James Van Allen of the State University of Iowa provided the instruments that detected the inner of Earth's two major trapped radiation belts that now bear his name.

As the United States turned its attention to the scientific exploration of outer space, an act of the U.S. Congress formed the National Aeronautics and Space Administration (NASA)—the civilian space agency, which opened its doors for business on October 1, 1958. Soon, Cape Canaveral thundered with the sound of rockets that aerospace engineers had converted from warhead-carrying military ballistic missiles to payload-carrying space launch vehicles.

In a magnificent wave of scientific exploration, NASA used the Cape to send progressively more sophisticated robot spacecraft to the Moon, around the Sun, and to all of the major planets in our solar system (even tiny Pluto). One epic journey, the *Voyager 2* mission, started from complex 41 when a mighty Titan/Centaur launch vehicle ascended flawlessly into the Florida sky on August 20, 1977. This hardy, robot explorer conducted a "grand tour" mission of all the giant outer planets (Jupiter, Saturn, Uranus, and Neptune). Then, like its twin (*Voyager 1*), the hardy robot spacecraft departed the solar system on an interstellar trajectory.

To date, four human-made objects (NASA's *Pioneer 10* and *11* spacecraft and the *Voyager 1* and *2* spacecraft), have achieved deep-space trajectories that will allow them to wander among the stars for countless millennia. Appropriately, each of these far-traveling spacecraft carries a message from Earth. *Pioneer 10* and *11* carry a special plaque, while *Voyager 1* and *2* bear a special recorded message. Of special significance here is that our first interstellar emissaries started their journeys through the galaxy from Cape Canaveral.

Spacecraft launched from the Cape have revolutionized astronomy, astrophysics, and cosmology. For example, NASA's four great astronomical observatories, the *Hubble Space Telescope (HST)*, the *Compton Gamma Ray Observatory (CGRO)*, the *Chandra X-ray Observatory (CXO)*, and the *Spitzer Infrared Telescope Facility (SIRTF)*, all reached outer space following successful rocket rides from the Cape. Data from these and many other scientific spacecraft have changed our understanding of the universe and revealed some of its most majestic and exciting phenomena.

Any celebration of the Cape's significance in history must also pay homage to the political decisions and technical efforts that allowed American astronauts to first walk on another world. Responding to President Kennedy's 1961 mandate, NASA focused the Mercury, Gemini, and Apollo Projects to fulfill his bold vision of sending American astronauts to the Moon and returning them safely to Earth in less than a decade. The Apollo Project, in particular, required the largest American rocket ever built—the colossal 363-foot- (111-m-) tall Saturn V rocket, which was the brainchild of Wernher von Braun.

However, Cape Canaveral Air Force Station, which had served the nation's space program so well, proved inadequate (too small) as a launch site for this monstrous vehicle. So, NASA acquired scrubland and marsh for buffer space to accommodate flight of this giant rocket. NASA called the new Saturn V launch sites Complex 39A and Complex 39B—it is the place from which human beings first ventured to another world. Following the Apollo Project, NASA modified Complex 39 to accommodate the space shuttle.

In this century, the Cape continues to provide launch services for numerous military, scientific, and commercial missions. Within the perspective of cosmic evolution, access to space from the Cape represents a grand manifestation of human intelligence reaching into the galaxy. Because of the Cape, human beings have successfully started on the pathway to the stars. There is no turning back. For humankind now, it is truly the universe or nothing.

VANDENBERG AIR FORCE BASE

Vandenberg Air Force Base is located 55 miles (89 km) north of Santa Barbara near Lompoc, California. It is the site of all military, NASA, and commercial space launches accomplished on the West Coast of the United States. These launches primarily support missions requiring polar orbits around Earth. The base, named in honor of General Hoyt S. Vandenberg (U.S. Air Force chief of staff from 1948 to 1953), also provides launch facilities for the testing of intercontinental ballistic missiles. The first missile was launched from Vandenberg in 1958 and the world's first polar-orbiting satellite was launched from this facility by a Thor/Agena launch vehicle in 1959.

WALLOPS FLIGHT FACILITY

The NASA Goddard Space Flight Center's Wallops Flight Facility (WFF) is located on the eastern shore of Virginia. The facility was originally established in 1945 by the National Advisory Committee for Aeronautics (NACA) as a center for aeronautic research. When NASA absorbed NACA in 1958, the site became Wallops Station. Then, on April 26, 1974, NASA renamed the site the Wallops Flight Center. After consolidating with NASA's Goddard Space Flight Center (GSFC) in 1982, the facility was renamed the Wallops Flight Facility.

As one of the world's oldest launch sites, WFF is now NASA's primary facility for management and implementation of suborbital research programs. Since 1945 this facility has launched more than 14,000 rockets—primarily sounding rockets. Over the years, the WFF launch range has grown to include six launch pads, assembly facilities and state-of-the-art instrumentation. In addition, the facility's mobile launch facilities enable NASA scientists and engineers to launch rockets around the world.

BAIKONUR COSMODROME

The Baikonur Cosmodrome is a major launch site for the space program of the former Soviet Union and now the Russian Federation. The complex is located just east of the Aral Sea on the barren steppes of Kazakhstan (an independent republic). Constructed in 1955 when Kazakhstan was an integral part of the former Soviet Union, this historic cosmodrome covers over 2,590 square miles (6,700 km^2) and extends some 47 miles (75 km) from north to south and 56 miles (90 km) east to west. The sprawling rocket base includes numerous launch sites, nine tracking stations, five tracking control centers, and a companion 930-mile- (1,500-km-) long rocket test range.

Also known as the Tyuratam launch site during the cold war, this complex was used by the Soviets to launch *Sputnik 1* (1957), the first artificial satellite, and cosmonaut Yuri Gagarin (1934–1968)—the first human to travel into outer space (1961). The Russians have also used Baikonur Cosmodrome as the launch site for all their human-crewed missions and for the vast majority of their lunar, planetary, and geostationary Earth orbit missions. Since 1993, the Russian Federation has rented Baikonur from Kazakhstan.

GUIANA SPACE CENTER

In 1964, the French government selected Kourou, French Guiana, as the site from which to launch its satellites. After the European Space Agency (ESA) was formed in 1975, the French government offered to share the Guiana Space Center, or Centre Spatial Guyanais, with ESA. For its part, ESA approved funding to upgrade the launch facilities at Kourou to prepare the spaceport for the family of Ariane launch vehicles then under development.

Since that time, ESA has continued to fund two-thirds of the spaceport's annual budget in order to finance the operations and facilities needed to accommodate an evolving family of European space launch vehicles, especially the Ariane family of rockets.

Kourou lies at approximately five degrees north latitude and 52.4 degrees west longitude—just 310 miles (500 km) above

A Soyuz–Fregat launch vehicle lifts off from the Baikonur cosmodrome on July 15, 2000. (European Space Agency/ Starsem [Copyright 2000 © European Space Agency; used with permission])

the equator on the Atlantic Ocean in coastal French Guiana. This location makes the space complex ideally situated for launches into geostationary transfer orbit. Thanks to its favorable geographical position in the northeast corner of South America, Europe's spaceport can also support a wide range of missions from due east (e.g., geostationary transfer orbit) to north (e.g., polar orbit). In fact, Kourou is so well placed that with just one spaceport all possible European space missions can be launched with minimum risk. Tropical forests largely cover French Guiana and the launch site itself has no significant natural risks from either hurricanes or

A panoramic view of the Guiana Space Center, Europe's spaceport, which is situated in the northeast corner of South America in Kourou, French Guiana (November 2004) *(European Space Agency [Copyright 2000–2004 © European Space Agency; used with permission])*

earthquakes. In addition to clients for space transportation services from countries throughout Europe, Kourou also provides launch services to aerospace industry clients from the United States, Japan, Canada, Brazil, and India.

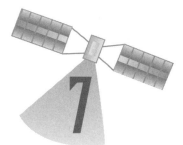

Upper-Stage Rockets and Orbital Transfer Vehicles

Often, getting a payload into low Earth orbit (LEO) is only part of the overall rocket propulsion effort. Some payloads need an upper-stage vehicle that thrusts them to a higher altitude orbit around Earth or places them on an Earth-escape trajectory into interplanetary space. This chapter describes some of the upper-stage rocket vehicles that American aerospace engineers have used to accomplish this important task. Included here are the Agena upper-stage rocket vehicle, the Centaur upper-stage rocket vehicle, and the inertial upper stage (IUS) system.

✧ Upper-Stage Vehicles

At present, aerospace engineers design upper-stage rocket vehicles with either liquid-propellant or solid propellant chemical engines—although nuclear thermal rocket engines, like NERVA, have also been considered. Upper-stage rocket engines must survive the rigors of launch from Earth's surface and then remain functional for some period of time in the space environment following separation from the booster rocket.

American aerospace engineers can use either solid- or liquid-propellant upper-stage rocket engines with expendable launch vehicles. However, NASA's human space flight safety guidelines require that all upper-stage rocket vehicles carried within the space shuttle's cargo bay be previously approved solid-propellant rocket engine systems.

Currently, all upper-stage rocket engines are expendable, one-time-use-only systems. The space tug is a proposed (but as yet undeveloped) re-usable upper-stage vehicle that would be capable of delivering, retrieving, and servicing payloads in orbits and trajectories beyond low Earth orbit (LEO). NASA's proposed space tug would be "hangared" in space (possibly near the permanent space station), refueled and refurbished on orbit, and reused many times.

✧ Orbital Transfer Vehicle

A propulsion system used to transfer a payload from one orbital location to another—as, for example, from low Earth orbit (LEO) to geostationary Earth orbit (GEO). Orbital transfer vehicles (OTVs) can be expendable or reusable; many involve chemical rocket systems, although solar electric propulsion systems have also been used and nuclear electric propulsion (NEP) systems are also being considered. An expendable orbital transfer vehicle is also called an upper-stage unit, while a reusable OTV is known as a space tug. OTVs can be designed to move people and cargo between different interplanetary destinations—as, for example, between orbital locations within cislunar space or from an orbit around Earth to an orbit around Mars.

One frequently used OTV is called the apogee kick motor (or apogee rocket). This system is a solid-propellant rocket motor that is attached

HOHMANN TRANSFER ORBIT

The Hohmann transfer orbit is the most efficient orbit transfer path between two coplanar circular orbits. The maneuver consists of two impulsive high-thrust burns (or firings) of a spacecraft's propulsion system. The first burn is designed to change the original circular orbit to an elliptical orbit whose perigee is tangent with the lower-altitude circular orbit and whose apogee is tangent with the higher-altitude circular orbit. After coasting for half of the elliptical transfer orbit and when tangent with the destination circular orbit, the second impulsive high-thrust burn is performed by the spacecraft's onboard propulsion system.

The onboard propulsion system is often called an apogee kick motor. This action circular-

izes the spacecraft's orbit at the desired new altitude. The technique can also be used to lower the altitude of a satellite from one circular orbit to another circular orbit (of lower altitude). In this case, two impulsive retrofirings are required. The first retrofire (that is, a rocket firing with the thrust directed opposite to the direction of travel) places the spacecraft on an elliptical transfer orbit. The second retrofire then takes place at the perigee of the elliptical transfer orbit; it circularizes the spacecraft's orbit at the desired lower altitude.

This important maneuver is named after Walter Hohmann (1880–1945), the German engineer who proposed the orbital transfer technique in the 1925 book *The Attainability of Celestial Bodies*.

to a spacecraft and fired when the deployed spacecraft is at the apogee of an initial (relatively low-altitude) parking orbit around Earth. The precise firing establishes a new orbit farther from Earth or permits the spacecraft to achieve escape velocity.

✧ Agena

The Agena was a versatile upper-stage rocket that supported numerous American military and civilian space missions in the 1960s and 1970s. One special feature of this liquid-propellant system was its in-space engine restart capability. The United States Air Force originally developed the Agena for use in combination with Thor or Atlas rocket first stages. Agena A, the first version of this upper stage, was followed by Agena B, which had a larger fuel capacity and engines that could restart in space. The Agena A was 19.5 feet (5.94 m) long, had a diameter of 5.0 feet (1.52 m), and produced a thrust of approximately 15,525 pounds-force (69,000 N) for 120 seconds. The later Agena D was designed to provide a launch vehicle for a variety of military and NASA payloads. The Agena B and Agena D versions were both 24.8 feet (7.56 m) long, 5.0 feet (1.52 m) in diameter, and produced a thrust of about 15,975 pounds-force (71,000 N) for 240 seconds.

NASA used Atlas-Agena vehicles to launch large Earth-orbiting satellites, as well as many lunar and interplanetary space probes. NASA used Thor-Agena vehicles to launch scientific satellites, such as the *Orbiting Geophysical Observatory* (OGO), and applications satellites, such as *Nimbus* meteorological satellites. In the Gemini Project, the Agena D vehicle, modified to suit specialized requirements of space rendezvous and docking maneuvers, became known as the Gemini Agena Target Vehicle (GATV).

✧ Centaur

The Centaur is powerful and versatile upper-stage rocket originally developed by the United States in the 1950s for use as a high-performance upper-stage rocket with the Atlas launch vehicle. The Centaur vehicle's rocket engine uses the energetic cryogenic-propellant combination of liquid hydrogen and liquid oxygen. It was the first American rocket to successfully burn liquid hydrogen as its fuel. Centaur has supported many important military and scientific missions, including NASA's Cassini spacecraft mission to Saturn, which was launched from Cape Canaveral on October 15, 1977 and successfully arrived at Saturn on July 1, 2004.

There are a number of versions of the Centaur upper-stage rocket. When used with the Atlas launch vehicle, the Centaur is typically 29.8 feet

On October 15, 1997, a Titan IV/Centaur rocket—at the time, the United States' most powerful expendable rocket vehicle combination—successfully lifted off from Cape Canaveral Air Force Station, Florida, and started the *Cassini* spacecraft (with attached *Huygens* probe) on its successful seven-year journey through the solar system. On July 1, 2004, the two-story-tall *Cassini* reached the magnificent ringed planet and began its four-year mission of intensive scientific investigation within the Saturn system. *(NASA and United States Air Force)*

(9.1 m) long, 10.0 feet (3.05 m) in diameter, and develops a thrust of about 29,925 pounds-force (133,000 N)—although longer versions (up to 38.3 feet [11.68 m]) develop a thrust of up to approximately 44,550 pounds-force (198,000 N). Aerospace engineers use the Centaur T version with the powerful Titan IV launch vehicle, the Centaur T is 29.5 feet (9.0 m) long, 14.1 feet (4.3 m) in diameter, and develops approximately 33,100 pounds-force (147,000 N) of thrust.

The Centaur vehicle is actually wider than the main (core) liquid-propellant engine of the Titan launch vehicle. So, when the Centaur sits on top of a modern Titan launch vehicle, it creates an unusual sight that aerospace engineers call the "hammerhead" configuration.

✧ Inertial Upper Stage

Inertial upper stage (IUS) is a versatile orbital transfer vehicle (OTV) developed by the United States Air Force and manufactured by the Boeing Company. The IUS is a two-stage payload delivery system that travels from low Earth orbit (LEO) to higher-altitude destinations, such as geosynchronous orbit (GEO). The versatile upper stage vehicle is compatible with both the space shuttle and the Titan IV expendable launch vehicle.

The IUS consists of two high-performance, solid-propellant rocket motors, an interstage section, and supporting guidance, navigation, and communications equipment. The first stage rocket motor typically contains about 21,350 pounds (9,700 kg) of solid propellant and generates a thrust

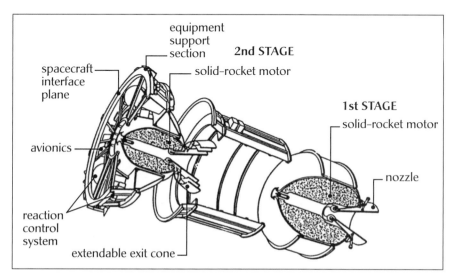

A cutaway drawing of the inertial upper stage (IUS) system, which shows all the major components of this versatile, solid-propellant orbital transfer vehicle *(NASA)*

of about 41,600 pounds-force (185,000 N). The second-stage solid-rocket motor contains about 6,000 pounds (2,720 kg) of propellant and generates some 17,600 pounds-force (78,300 N) of thrust. The nozzle of the second stage has an extendable exit cone for increased system performance.

In July 1999, an IUS system propelled NASA's *Chandra X-ray Observatory* from LEO into its operational highly elliptical orbit around Earth that reaches one-third of the distance to the Moon. Furthermore, the IUS has helped send the *Galileo* spacecraft on its journey to Jupiter, the *Magellan* spacecraft to Venus, and the *Ulysses* spacecraft on its extended journey to explore the polar regions of the Sun. The U.S. Air Force has used IUS rockets to send early warning satellites, such as the large Defense Support Program (DSP) spacecraft, from low Earth orbit to their operational locations in geosynchronous orbit.

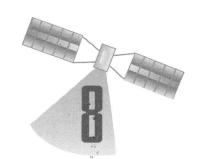

Sounding Rockets

This chapter discusses the basic functions of the sounding rocket and the characteristics of some of the more popular systems that have been used in the past six decades.

A sounding rocket is a rocket used to carry scientific instruments on parabolic trajectories into the upper regions of Earth's sensible atmosphere (i.e., beyond the reach of aircraft and scientific balloons) and into near-Earth space. The sounding rocket is basically divided into two major components: the rocket propulsion system (usually with a solid-propellant motor), and the payload section. Sounding rocket payloads typically include: the experiment package, the nose cone, a telemetry system, an attitude control system, a radar-tracking beacon, the firing despin module, and the recovery section.

In many cases, sounding rocket payloads must operate without the residual spinning motion imparted by the launch vehicle. In other special cases, payloads must have very specific spin rates in order to accomplish the desired scientific goals. Aerospace engineers accomplish payload despin (to zero revolutions per minute or to a predetermined spin rate) by using mechanical yo-yo despin systems that release masses attached to flyaway cables. These devices are wrapped around the payload compartment's circumference and unwind when released. This despin technique is relatively simple, yet has proven very reliable. Rocket engineers often use the despin system to despin both the launch vehicle's final stage as well as the payload prior to payload separation. However, during some missions, the engineers despin just the payload following its separation from the spent rocket vehicle.

The primary use of the recovery system is to retrieve the payload after a sounding rocket flight so it can be refurbished and flown again. The NASA sounding rocket program uses two basic types of recovery system: land and water. The land recovery system uses various sized parachutes (depending on payload mass) to achieve an acceptably "soft" impact of the payload on land. As the payload descends back to Earth on a ballistic trajectory at the end of a sounding rocket flight, the parachute recovery system typically deploys in the 16,400-foot (5,000-m) to 21,300-foot (6,500-m) altitude region. A variety of recovery aids, such as a homing beacon transmitter, flashing strobe light, and smoke or dye marker, assist search personnel in locating and recovering the payload. The water recovery system used on NASA sounding rockets can float a payload (up to 100 kg in mass) that splashes down in a body of water. For sounding rockets launched from NASA's Wallops Flight Facility, water recovery takes place in the Atlantic Ocean and is accomplished by boat (U.S. Coast Guard, U.S. Navy, or commercial source) or by helicopter.

A four-stage Black Brant XXII sounding rocket blasts off its launch pad at NASA's Wallops Island Flight Facility in the coastal regions of rural Virginia (January 1988). *(NASA)*

✧ The Modern Sounding Rocket

Modern sounding rockets fly vertical trajectories that range from 31 miles (50 km) to more than 780 miles (1,250 km) in altitude. A sounding rocket flight normally lasts less than 30 minutes. Sounding rockets come in a wide variety of sizes and types. For example, NASA's current sounding rocket fleet ranges from the single-stage Super Arcas, which stands 9.8 feet (3 m) high and 4.3 inches (11 cm) in diameter, to the single-stage Aries, which stands 36 feet (11 m) high and 3.6 feet (1.1 m) in diameter.

NASA has used the Super Arcas since 1962 for carrying meteorological measuring devices. This sounding rocket can carry a 10 pound-mass (4.5 kg) payload to an altitude of 56 miles (90 km). The four-stage Black Brant XII, the tallest in NASA's fleet, is 66 feet (20 m) high. The performance characteristics of several of these sounding rockets are described near the end of this chapter.

Researchers like to use sounding rockets because these devices provide a relatively inexpensive, yet generally reliable, way to conduct research in otherwise unreachable portions of Earth's upper atmosphere. Rocket scientists derived the term *sounding rocket* by making an analogy to "maritime soundings," which involve scientific instruments placed at various depths in the world's oceans.

NASA's current collection of sounding rockets uses solid-propellant rocket motors. To keep costs low, the majority of these sounding rocket systems also make extensive use of 20- to 30-year-old military surplus rocket motors—especially the basic Nike solid-propellant rocket motor, which produces an average sea-level thrust of approximately 42,800 pounds-force (190,300 N). All of these rocket vehicles are unguided, except the Aries and those that use the S-19 Boost Guidance System. The S-19 Swedish (SAAB)–developed Boost Guidance System is currently used by NASA on the Black Brant V, Nike-Brant V, Terrier-Black Brant V, and Black Brant X sounding rockets. The main purpose of the S-19 guidance system is to reduce impact dispersion and to acquire accurate trajectories. With this guidance system, it is possible to launch sounding rockets to higher altitudes and in higher winds, while maintaining tolerable trajectory dispersion.

During flight, all sounding rockets (except the Aries) are imparted with a spinning motion to reduce potential dispersion of the flight trajectory due to vehicle misalignments. NASA conducts between 30 to 35 suborbital sounding rocket flights per year, providing low-cost, quick-response flight opportunities to scientists in the areas of physics and astronomy, microgravity materials processing research, and instrument development. The majority of NASA's sounding rocket launches currently take place at Wallops Flight Facility on Wallops Island, Virginia. However, NASA and foreign space research agencies and organizations also launch sounding rockets from a variety of locations around the world.

The versatility and mobility of modern sounding rockets has permitted scientists to use them in numerous research applications from both temporary and permanent (fixed) launch sites around the world—including, and especially in, remote regions that have special scientific interest. Sounding rockets are also launched from ships at sea. One interesting sounding rocket variant is called the "rockoon." The rockoon is a high-altitude sounding system consisting of a small solid-propellant rocket carried aloft by a large (helium-filled) plastic balloon. The rocket is fired near the maximum altitude of the balloon flight. The rockoon represents a relatively mobile atmospheric sounding system and has been used extensively from ships.

✦ Sounding Rockets Sites around the World

From above the Arctic Circle to below the Antarctic Circle and at many remote locations in between, scientists have launched sounding rockets in pursuit of new knowledge about Earth's upper atmosphere and geophysical phenomena in near-Earth space. In the early 1960s, for example, the Andøya Rocket Range in northern Norway (at approximately 69.3°N latitude and 16.0°E longitude) was established and initially used to launch American sounding rockets, including the NASA's Nike Cajun rocket. Today, the Norwegian Space Center operates this polar region rocket range (consisting of eight launch pads) for a variety of commercial and scientific customers.

The Esrange launch site in northern Sweden (at approximately 67.9°N latitude and 21.1°E longitude) is another example of a permanent (fixed) facility in a remote location that launches sounding rockets in Earth's polar regions. The Esrange site serves as the space operation center of the Swedish Space Corporation (SSC). Initially established in March 1964 by the European Space Research Organization—the forerunner to the European Space Agency—the SSC has owned and operated the facility since 1972. About 125 miles (200 km) north of the Arctic Circle, Esrange is located in the auroral zone, which makes it perfect for the investigation of high-latitude geophysical phenomena and auroral studies. The Esrange space research range has six permanent launchers and support facilities, including a blockhouse. The range is equipped with launchers for many different types of sounding rockets. The range's ground instrumentation permits two simultaneous launchings, or launchings of several rockets in rapid succession. Esrange currently provides sounding rocket launch services for scientists from all over the world—especially those from member nations of the European Space Agency. A cooperative agreement exists between Esrange and the Norwegian sounding rocket range at Andøya, Norway.

But the sounding rocket is not a "space-age" idea. Sending scientific instruments into the upper regions of Earth's atmosphere was one of the principal motives for modern rocket development at the start of the 20th century. In 1914, the American rocket scientist Robert Goddard was one of the first to identify the important relationship between the rocket and upper atmospheric research. Throughout his career, he made the scientific sounding rocket a major objective in his pioneering rocket design studies. In September 1916, Goddard sent a description of his early rocket experiments along with a modest request for funding to the Smithsonian Institution. Upon reviewing Goddard's proposal, officials at the Smithsonian found his rocket experiments to be "sound and ingenious." One key point in Goddard's favor was his well-expressed desire to develop the liquid-propellant rocket as a means of taking instruments into the high-altitude regions of Earth's atmosphere—regions that were well beyond the reach of weather balloons. This early marriage between rocketry and meteorology proved very important, because the relationship provided officials in funding institutions a tangible reason for supporting Goddard's proposed rocket work.

Not only did the Smithsonian provide additional funding to Goddard over the years, but the organization also published Goddard's two classic monographs on rocket propulsion: the first in 1919 and the second in 1936. In 1919, Goddard summarized his early rocket theory work in the classic report "A Method of Reaching Extreme Altitudes." This was the first American scientific work that carefully discussed all the fundamental principles of rocket propulsion. Goddard described the results of his experiments with solid-propellant rockets and even included a final chapter on how the rocket might be used to get a modest payload to the Moon.

Goddard also conducted a series of important liquid-propellant rocket experiments the 1930s in a remote part of the New Mexico desert, near Roswell. In rural New Mexico, he was able to test his rockets well out of the public view and so avoided the painfully adverse publicity that had accompanied his previous rocket testing efforts in the more densely populated regions of the northeastern United States. These rocket experiments in New Mexico led to Goddard's second important Smithsonian monograph, *Liquid-Propellant Rocket Development,* published in 1936. On March 26, 1937, Goddard launched a liquid-propellant rocket, nicknamed "Nell" (as were all his flying rockets), which flew for 22.3 seconds and reached a maximum altitude of about 8,860 feet (2,700 m). This was the highest altitude ever obtained by one of Goddard's rockets. Despite the modest altitude by today's sounding rocket standards, Goddard's pioneering work led the way for many technical advances in modern rocketry—important advances applicable to all aspects of rocketry, not just sounding rockets.

However, the true significance of Goddard's rocket research went essentially unnoticed by officials within the U.S. government, because they regarded the rocket as a technical novelty. In fact, it was not until 1945 (the year in which Goddard died) that the first U.S. government-sponsored sounding rocket was launched. That early sounding rocket, called the WAC Corporal, flew in October 1945 as part of a collaborative project between the Jet Propulsion Laboratory of the California Institute of Technology and the ordnance department of the army. It was also the first rocket launched from the White Sands Missile Range.

The WAC Corporal sounding rocket carried a 24-pound (11-kg) package of meteorological instruments as its payload and rose to an altitude of about 25 miles (40 km). This early sounding rocket used a Tiny Tim (solid-propellant) rocket to accelerate out of the tower at Launch Complex 33. The U.S. Army's liquid-propellant Corporal rocket provided the propulsion technology for this small, high-altitude sounding rocket. Although the source of the name "Corporal" is apparent, the precise origin of "WAC" is not quite as clear. Some rocketry historians suggest that WAC is an acronym for "without attitude control." This appears logical, since the early sounding rocket did not have an onboard guidance or stabilization system and relied on its three tail fins for in-flight stabilization. These three fins

On February 24, 1949, a Bumper-WAC rocket was launched from the U.S. Army's White Sands Proving Grounds (now called the White Sands Missile Range) and reached a record-setting altitude of approximately 400 kilometers. The Bumper-WAC vehicle consisted of a captured and refurbished German V-2 rocket and a U.S. Army–sponsored WAC-Corporal second stage. The Bumper-WAC rocket vehicle was the world's first large two-stage liquid-propellant rocket. *(U.S. Army/ White Sands Missile Range)*

deployed after the high altitude research rocket cleared the launch tower. Engineers designed the nose cone of the WAC Corporal rocket to separate near the end of its flight and to release a parachute so that the package of meteorological instruments it carried could be recovered.

Over the years, a variety of interesting sounding rockets have collected leading-edge data that stimulated scientific progress in a number of fields, including meteorology, astronomy, astrophysics, national-defense related missile and sensor research, geophysics, auroral research, galactic astronomy, and particle physics.

Sounding rockets have supported routine atmospheric measurements at altitudes as low as 22 miles (35 km)—an important region of Earth's atmosphere that cannot be directly sampled by aircraft or spacecraft. The use of sounding rockets has also allowed scientists to make measurements at altitudes up to 4,000 miles (6,400 km)—accomplishing in-situ investigations of the ionosphere, conducting near-Earth space physics research, and supporting pioneering efforts in X-ray astronomy, high-energy astrophysics, and gamma-ray astronomy.

On February 24, 1949, a high-altitude rocket vehicle called Bumper-WAC was launched from the U.S. Army's White Sands Proving Ground—now called the White Sands Missile Range (WSMR). The rocket vehicle reached a record altitude of 250 miles (400 km) and became the first known human-made object to reach outer space.

This altitude record lasted until 1957—the year that witnessed the start of the space age. The first stage of the Bumper-WAC rocket vehicle was a captured and refurbished German V-2 (A-4) rocket that had its warhead replaced by a launching compartment. After the V-2's rocket engines shut down, there was a noticeable high-altitude "bump." According to aerospace folklore, this "bump" inspired rocket engineers to give the name "Bumper" to the new rocket vehicle configuration. The second stage of the Bumper-WAC rocket vehicle was a modified WAC Corporal sounding rocket that had been mounted in the nose cone. The Jet Propulsion Laboratory was responsible for the design study, engineering development, and testing of the Bumper-WAC vehicle in collaboration with the General Electric Company and the Douglas Aircraft Company. The overall effort was supervised and sponsored by the U.S. Army's ordnance department. The high-altitude Bumper-WAC vehicle was the world's first large two-stage liquid-propellant rocket system.

In 1949, sounding rocket–borne instruments also performed the first solar X-ray measurements. Then, some 13 years later in June 1962, Professor Bruno B. Rossi (1905–93) and his associates placed a special instrument package on a sounding rocket flight and unexpectedly detected the first nonsolar source of cosmic X-rays, called *Scorpius X-1*. This event is often regarded as the start of X-ray astronomy.

WHERE DOES OUTER SPACE BEGIN?

The United Nations Committee on the Peaceful Uses of Outer Space (COPUOS) has been and continues to be the main architect of international space law; it was established by resolution of the U.N. General Assembly in 1953 to study the problems associated with the arrival of the space age. COPUOS is made up of two subcommittees, one of which studies the technical and scientific aspects and the other the legal aspects of space activities. One contemporary space law topic now under international discussion is the delimitation of outer space—that is, where does outer space begin and national "air-space" end from a legal point of view? As used in this book and as also frequently encountered in the technical literature of the aerospace field, outer space is a general term that refers to any region beyond Earth's atmospheric envelope. By informal international agreement, scientists and engineers generally consider outer space to begin somewhere between 62 miles (100 km) and 125 miles (200 km) altitude. However, within the U.S. human spaceflight program, persons who have traveled beyond 50 miles (80 km) altitude are usually recognized as space travelers or astronauts.

Less than a decade later, sounding rockets played an important role in the International Geophysical Year (IGY)—an 18-month period of international scientific investigation (from July 1, 1957 to December 31, 1958) that coincided with a high level of solar activity. The IGY promoted an intensive study of the natural environment—Earth, its oceans and atmosphere. The global effort involved some 30,000 scientific participants representing 66 countries. As part of this ambitious scientific program, IGY scientists launched more than 300 instrumented-sounding rockets from sites around the world. Data collected by these sounding rocket flights contributed to significant discoveries regarding the atmosphere, the ionosphere, auroras, geomagnetism, and the sources of cosmic radiation. However, the important role played by the sounding rocket during the IGY was somewhat overshadowed by the launch of *Sputnik 1* (the world's first artificial satellite) on October 4, 1957, and then, on January 31, 1958, the launch of *Explorer 1* (the first American satellite, which discovered Earth's artificial radiation belts).

In the 1960s, researchers in the Department of Defense launched a total of 141 four-stage solid-propellant Athena sounding rockets from a site in Green River, Utah. These tests were jointly conducted by personnel from the U.S. Army and the U.S. Air Force and used to simulate the flight environment of intercontinental ballistic missiles. During each test, the Athena rocket's payload, generally a specially designed reentry vehicle, would travel some 470 miles (750 km) away from the

A four-stage, solid–propellant Athena sounding rocket on its launcher in Green River, Utah (July 1966). Following launch, this rocket's payload, a test reentry vehicle, impacted on the White Sands Missile Range in New Mexico. *(U.S. Army/ White Sands Missile Range)*

launch point in Utah and impact on the White Sands Missile Range in New Mexico.

The International Quiet Sun Years (IQSY) was a full-scale follow-up to the IGY and took place between January 1, 1964 and December 31, 1965. IQSY was an intensive effort of geophysical observations during a period of minimum solar activity. Instrumented sounding rockets again played a significant role in the scientific investigation of Earth-Sun interactions. Scientists from many nations launched sounding rockets in cooperative research efforts from ranges in the United States and around the world. In addition to an investigation of Earth-Sun interactions, sounding rocket research during this period gave rise to three new branches of astronomy—ultraviolet, X-ray, and gamma ray. Experiments launched on sounding rockets contributed to an accurate characterization of Earth's upper atmosphere and contributed to knowledge of the chemistry of the ionosphere, enabled the detection of electrical currents in the ionosphere, and provided accurate descriptions of the particle fluxes in auroras. At the beginning of the space age, scientists also used sounding rockets to test new instruments that were planned for use on scientific spacecraft. That testing practice is still followed today. Furthermore, scientists use the quick response capabilities of sounding rockets to investigate transient geophysical phenomena, as well as unanticipated, but very exciting, astrophysical events—such as the great naked-eye supernova of 1987 (called supernova 1987A).

Because higher performance sounding rockets were not cost-effective for low-altitude experiments and lower-performance (yet less-expensive) sounding rockets were not useful for high-altitude experiments, NASA has used a number of sounding rockets of varying capabilities. Several well-known sounding rocket systems, and their main characteristics, are given below (in alphabetical order).

✧ Aerobee and Astrobee

Development of the Aerobee liquid-propellant sounding rocket started in 1946, when the U.S. Navy gave a contract to the Aerojet Engineering Corporation (later called the Aerojet-General Corporation) for the development of a high-altitude sounding rocket. The Applied Physics Laboratory (APL) at Johns Hopkins University was assigned technical direction of this project. James Van Allen (who would discover Earth's trapped radiation belts in 1958) served as the director of the Aerobee Project at APL. He came up with the sounding rocket's name by combining "Aero" from Aerojet Engineering and "bee" from "Bumblebee," the U.S. Navy's code name for its overall post–World War II project to develop naval rockets.

The initial Aerobee system was a single-stage liquid-propellant rocket, which used a solid-propellant rocket motor as its booster. This sounding rocket configuration needed a tall tower to guide the rocket-propelled vehicle while solid-propellant booster burned. Once the solid-propellant booster motor burned out (in about 2.5 seconds), the slowly accelerating Aerobee cleared the tower, jettisoned its booster, ignited its liquid-propellant engine, and deployed fins to guide it dur-

ing flight. The Aerobee's liquid propel-lant rocket engine used aniline as the fuel and red fuming nitric acid as the oxidizer. Aniline is a toxic, oily, colorless chemical propellant derived from benzene. It has a boiling point of 363°F (184°C) and its melting point is 21°F (−6°C). This innova-tive liquid-propellant engine developed a thrust of approximately 4,050 pounds-force (18,000 N). The original Aerobee rocket vehicle was capable of lifting a 150-pound (68-kg) payload to a nominal altitude of 80 miles (130 km). The term *nominal altitude* means a sea-level launch at an 85-degree launch elevation. The Aerobee was the first American general purpose high-altitude sounding rocket. Between 1947 and 1985, the military services and (later) NASA made extensive use of Aerobee and the Astrobee sounding rockets.

Once the basic Aerobee sounding rocket became available, it quickly found use in all the United States military services, as well as in within the newly established civilian space agency, NASA. However, to meet growing demands for improved high-altitude perfor-mance and payload capacity, a large number of variants—including the Aerobee 150, the Aerobee 170, the Aerobee 200, and Aerobee-Hi—soon appeared. In 1952, for example, at the request of the U.S. Air Force and the U.S. Navy, Aerojet Corporation undertook design and development of the Aerobee-Hi—a high-performance version of the Aer-obee sounding rocket engineered expressly for research in the upper atmosphere. One version of the Aerobee-Hi became the Aer-obee 150 system. The Aerobee 150 could take a 150-pound (68-kg) payload to a nominal altitude of 170 miles (270 km) or a 500-pound (227-kg) payload to a nominal

Liftoff of an Aerobee 150 sounding rocket from Launch Complex 35 at the White Sands Missile Range on January 17, 1985. (Note the tall tower needed to support this launch operation.) This was the last launch of the versatile and extensively used Aerobee sounding rocket. *(U.S. Army/ White Sands Missile Range)*

altitude of 70 miles (112 km). On April 30, 1957, an Aerobee Hi rocket, launched from the White Sands Missile Range, set a new altitude record for single stage sounding rockets when it carried a 155-pound (70-kg) payload to a height of 188 miles (302 km).

However, despite its capabilities, there were two major problems with the original Aerobee sounding rocket that limited its use in remote areas. First, it had a liquid-propellant sustainer rocket engine, which used the toxic propellants aniline for fuel and red fuming nitric acid as oxidizer. Second, this sounding rocket vehicle required a very tall launch tower, so the solid rocket motor booster could be guided as it burned and initially accelerated the Aerobee. To support remote operations and rapid firing of several sounding rockets in a salvo-like research investigation, Aerojet Corporation again improved the Aerobee by switching to a solid-propellant rocket motor for the vehicle's sustainer engine. The uprated Aerobee 150 was subsequently named the "Astrobee." With that important modification Aerojet Corporation started using the prefix "Aero" to designate liquid-propellant sounding rockets and "Astro" for its solid-fueled sounding rockets.

The Aerobee 170 was capable of carrying a 250-pound (114-kg) payload to a nominal altitude of 155 miles (250 km) or a 500-pound (227-kg) payload to an altitude of 93 miles (150 km). The Aerobee 170 earned the nickname "Workhorse of the Upper Atmosphere" because scientists so frequently used this research rocket as they probed near Earth space in the 1950s and 1960s.

The Aerobee 200 could carry a 250-pound (114-kg) payload to a nominal altitude of 193 miles (310 km) or a 500-pound (227-kg) payload to a nominal altitude of 180 miles (290 km). The Aerobee 350 was capable of carrying a 300-pound payload (136 kg) to a nominal altitude of 250 miles (400 km) or a 1,000-pound (454-kg) payload to a nominal altitude of 130 miles (210 km). Finally, the Astrobee 1500 was capable of carrying a 100-pound (45-kg) payload to a nominal altitude of 1,370 miles (2,200 km) or a 300-pound (136-kg) payload to a nominal altitude of 745 miles (1,200 km).

✧ Super Arcas

Scientists have used the Super Arcas rocket vehicle since 1962 for carrying meteorological measuring devices as high as 62 miles (100 km). This sounding rocket system is also used to take other types of scientific measurements in the same altitude region with little expense and rapid deployment. The Atlantic Research Corporation manufactures the vehicle's small solid-rocket motor.

Sounding Rockets 127

Arcas is an acronym for "All-purpose Rocket for Collecting Atmospheric Soundings." This acronym was an intentional choice in 1959 because the first three letters also corresponded to the initials of the Atlantic Research Corporation. The basic Arcas was a small solid-propellant sounding rocket developed specifically to support meteorological research. The original Arcas could carry a 11-pound (5-kg) payload to an altitude of 40 miles (64 km). Later versions were called the Boosted Arcas, the Boosted Arcas II, and the Super Arcas.

NASA currently uses the Super Arcas sounding rocket, which has successfully launched payloads ranging from 7.9 pounds (3.6 kg) to 18.3 pounds (8.3 kg). The Super Arcas solid propellant motor develops an average thrust of 325 pounds-force (1,450 N) and operates for a total of about 40 seconds. This motor, which has a mass of 83 pounds (37.7 kg) before launch, can typically lift a 10-pound (4.5-kg) payload to an altitude of 57 miles (92 km), when launched from sea level at an effective launch angle of 86 degrees. The Super Arcas rocket and its payload generally experience an acceleration of seven g's and a burnout roll rate of 25 revolutions per second.

(Scientists and engineers use the symbol "g" for the acceleration due to gravity. For example, at sea level on Earth the acceleration due to gravity is approximately 32.2 ft/s^2 [9.8 m/s^2]—a quantity scientists refer to as "one g." In aerospace engineering, this term is also used as a unit of stress for bodies experiencing acceleration. When a rocket vehicle is accelerated during launch, everything inside it [including the payload] experiences a force that may be as high as several g's.)

The payload section for a Super Arcas rocket usually has an outside diameter of 4.5 inches (11.4 cm) and a launch length of about 30 inches (76 cm). The payload section normally consists of a tangent ogive nose cone housing the experiment mated to a parachute recovery system.

Meteorologists use the term *rocketsonde* to describe a rocket-borne instrument package for the measurement and transmission of upper-air meteorological data (up to an altitude of about 47 miles (76 km), especially that portion of the atmosphere inaccessible to radiosonde techniques. Therefore, the Super Arcas frequently serves as a rocketsonde or meteorological rocket.

✧ Aries

In 1974, NASA participated with the Naval Research Laboratory, Sandia National Laboratories, and other organizations in developing a new sounding rocket, Aries. This vehicle used surplus solid-rocket motor second stages from U.S. Air Force Minuteman intercontinental ballistic missiles. As befitting its intended use in rocket-borne astronomical research, the rocket was named after a sign of the zodiac, Aries (the ram). The Aries rocket could carry 400-pound (180-kg) to 1,980-pound (900-kg) scientific payloads to altitudes that provided seven to 11 minutes of astronomical viewing time above an altitude of 62 miles (100 km), which space scientists consider as the nominal threshold of outer space. The Aries also had sufficient capability to provide between eight and 11 minutes of "weightless" conditions for payloads involving materials-processing experiments that needed microgravity environments.

✧ Black Brant

In the late 1930s, Bristol Aerospace Ltd. of Canada developed the Black Brant series of sounding rockets in cooperation with the Canadian government. The Canadian Armament Research and Development Establishment (CARDE) chose the name "Black Brant" for this research rocket—taking the name of a small, dark, fast-flying goose common to the northwest coast and Arctic regions of Canada. Since the first Black Brant rocket flew in 1939, the versatile vehicle has evolved into a family of sounding rockets used by scientists in research agencies around the world, including NASA. Today, the Black Brant is often used in combination with another, more powerful, solid-rocket motor, which serves as the combination vehicle's booster. As the Black Brant evolved, the Canadian government retained the rocket's original name and simply added Roman numerals to designate different members of the series. NASA started using Black Brant sounding rockets in 1970 and currently uses the Black Brant V, Nike-Black Brant V, Black Brant X, Black Brant IX, and Black Brant XII vehicles.

The Black Brant V is a single-stage solid-propellant sounding rocket. There is a three-fin version (designated VB) and a four-fin version (designated VC). The Black Brant V solid rocket motor operates for about 27 seconds and produces an average thrust level of 17,020 pounds-force (75,700 N). The primary diameter of the Black Brant V is approximately 1.44 feet (0.44 m), and the rocket has a length of 17.5 feet (5.33 m). The loaded mass of the Black Brant V rocket motor (including supporting hardware) is 2,800 pounds (1,270 kg), including 2,200 pounds (1,000 kg) of propellant. As an example of its capabilities, the Black Brant VB rocket can lift a 500-pound (225-kg) payload to a nominal altitude of about 155 miles (250 km), with the total parabolic flight taking about 490 seconds—a little over eight minutes.

Rocket engineers have improved the performance capabilities of the Black Brant V vehicle by mating it with a Nike booster rocket motor. The new configuration is called the Nike-Black Brant V B/C launch vehicle.

The Nike solid-propellant booster rocket is an adaptation of the Nike Ajax antiaircraft missile—developed in the late 1940s by the Hercules Powder Company for the U.S. Army. (In ancient Greek mythology, Nike was the winged goddess of victory.) The average sea-level thrust of the Nike rocket motor is approximately 42,800 pounds-force (190,300 N). The basic Nike motor is 11.3 feet (3.45 m) long with a principal diameter of 1.4 feet (0.42 m). To accommodate the Black Brant V, rocket engineers bolt an interstage adapter to the front of the Nike solid-propellant motor. The interstage adapter consists of a conical shaped adapter, which slip-fits into the second stage nozzle, thereby providing for drag separation when the Nike rocket motor experiences burnout. Normally, the Nike motor's fins are canted to provide a two-revolutions-per-second spin rate at booster burnout. The total mass of the Nike booster system (with supporting case hardware) is 1,320 pounds (600 kg), including 760 pounds (345 kg) of propellant.

The Nike and Black Brant V stages drag separate at Nike motor burnout and then the Black Brant V solid rocket ignites at 8.5 seconds flight time. The Nike-Black Brant VB sounding rocket configuration can take a 500-pound (225-kg) payload to a nominal altitude of 230 miles (370 km).

The Black Brant X launch vehicle is a three-stage sounding rocket system. It is somewhat unique because the third-stage motor, called the Nihka solid-propellant motor, ignites only after the rocket vehicle reaches exoatmospheric conditions. (Generally, multistage sounding rockets have all their rocket engine stages burn out in the lower atmosphere.) The first-stage rocket booster for the Black Brant X vehicle is a Terrier rocket motor with four external fin panels arranged in a cruciform configuration. The second stage is a Black Brant V rocket motor. The third stage is the Nihka rocket motor, which was developed by Bristol Aerospace Ltd. specifically for the Black Brant X system. The Nikha motor has a total mass of 890 pounds (405 kg) (including 760 pounds [345 kg] of propellant) and develops 12,000 pounds-force (53,400 N) of thrust. The standard payload configuration for the Black Brant X vehicle is 17.2 inches (43.8 cm) in diameter with a 3:1 ogive nose shape. For a launch elevation of 85 degrees, the Black Brant X vehicle can carry a 155-pound (70-kg) payload to an altitude of more than 775 miles (1,250 km).

The Black Brant XII is a four-stage sounding rocket system used primarily to carry a variety of scientific payloads to very high altitudes. This vehicle is a development spin-off of the Black Brant X system. The first stage of the Black Brant XII vehicle is a Talos solid rocket motor and the second stage is a Taurus solid rocket motor. The third stage is a modified Black Brant VC solid motor, which has the nozzle extended for exoatmospheric use. The fourth stage of this vehicle is a Nihka rocket motor. The standard payload configuration for the Black Brant XII vehicle is 17.2

inches (43.8 cm) in diameter with a 3:1 ogive nose shape. As an example of its capability, the Black Brant XII vehicle can take a payload with a mass of 310 pounds (140 kg) to an altitude of 930 miles (1,500 km). If the payload has a mass of about 1,155 pounds (525 kg), the Black Brant XII can lift it to a nominal altitude of about 310 miles (500 km) above sea level. With a total vehicle length of about 66 feet (20 m) the Black Brant XII is NASA's tallest sounding rocket.

Finally, NASA uses the Black Brant XI sounding rocket system to send heavy scientific payloads to high altitudes. This three-stage vehicle is a direct development spin-off of the Black Brant XII vehicle, since the Black Brant XI's solid rocket motor configuration consists of the Black Brant XII's first three stages—a Talos rocket motor, a Taurus rocket motor, and a modified Black Brant VC motor, respectively. The Black Brant XI vehicle can lift a 1,210-pound (550-kg) payload to a nominal altitude of 220 miles (350 km) and a 700-pound (320-kg) payload to a nominal altitude of approximately 310 miles (500 km).

Nuclear Rockets to Mars and Beyond

This chapter describes the physical principles behind the operation of the nuclear thermal rocket, provides a brief history of the U.S. nuclear rocket program as undertaken between 1955 and 1973, and examines some of the future applications of nuclear thermal rockets possible in this century—especially in support of human expeditions to Mars.

In a nuclear thermal rocket, chemical combustion of a fuel and oxidizer is not required. Instead, a single propellant, usually hydrogen (H_2), is heated by the energy released in the nuclear fission process, which occurs in a controlled manner in the reactor's core. As shown in the figure at the top of page 132, the nuclear thermal rocket uses a nuclear reactor to heat hydrogen to extremely high temperatures before it is expelled through a rocket nozzle. Conventional rockets, in which chemical fuels are burned, have severe limitations in the specific impulse (I_{sp}) that a given propellant combination can produce. For example, a liquid-propellant chemical rocket engine, which uses liquid hydrogen (LH_2) as its fuel and liquid oxygen (LOX) as its oxidizer, has a theoretical specific value of between 300 and 380 seconds in the American engineering unit system and between 2,940 and 3,725 meters per second in the international system (SI) of units. These performance limitations are imposed by the relatively high molecular weight of the combustion products.

The principles of rocket engineering and compressible fluid dynamics predict that specific impulse (and therefore rocket engine performance) increases as the combustion chamber temperature increases and decreases with the molecular weight of the exhaust gases. Consequently, aerospace engineers search for rocket engines that can heat the expelled gases to as high a temperature as possible and that can expel combustion product gases with as low a molecular weight as possible.

At attainable combustion chamber temperatures, the very best chemical rockets are limited to specific impulse values of about 440 seconds

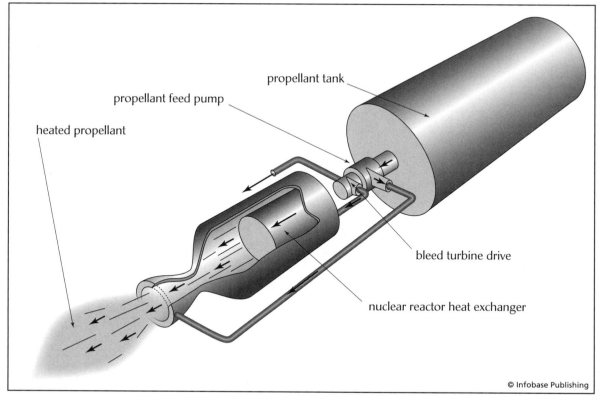

Basic components and principle of operation of a nuclear thermal rocket. The propellant tank carries liquid hydrogen at cryogenic temperatures. This single propellant is heated to extremely high temperatures as it passes through the nuclear reactor, and then very hot gaseous hydrogen is expelled out of the rocket's nozzle to produce thrust.

(4,300 meters per second in SI units). While searching for ways to build more powerful rocket engines in the late 1940s and early 1950s, engineers quickly recognized that nuclear thermal rocket systems using fission reactions would provide much greater propulsion performance capabilities. Some visionary rocket scientists even suggested the possibility that fusion (thermonuclear) reactions or matter-antimatter annihilation reactions could be harnessed in some type of advanced reaction engine to achieve rocket propulsion. This chapter focuses on nuclear thermal rockets that use fission reactors as the energy source.

✧ Nuclear Thermal Rocket Engineering

In nuclear fission, the nucleus of a special heavy element isotope, such as uranium-235, is bombarded by a neutron, which it absorbs. The resulting compound nucleus is unstable and soon breaks apart, or fissions, forming two lighter nuclei (called fission products) and releasing additional neutrons. In a properly designed nuclear reactor, these fission neutrons are used to sustain the fission process in a controlled chain reaction. The nuclear fission process is accompanied by the release of a very large amount of energy, typically 200 million electron volts (MeV) per reaction. By way of comparison, the most energetic chemical reactions liberate some 20 to 30 electron volts per reaction—about a million times less.

An electron volt (eV) is a tiny amount of energy equivalent to 1.519 $\times 10^{-22}$ British thermal unit (Btu) (1.6×10^{-19}J). Scientists and engineers almost universally use metric multiples of the electron volt (such as million electron volts [MeV]), when they describe processes and reaction energies at the atomic or nuclear levels. This book follows that customary practice.

Much of the energy liberated in nuclear fission appears as the kinetic (or motion) energy of the fission-product nuclei, which is then converted to thermal energy (or heat) as the fission products slow down in the reactor fuel material. In a nuclear thermal rocket, this thermal energy is removed from the reactor core by the flow of a fluid (such as hydrogen), which then becomes a very hot propellant.

This intensely hot hydrogen flows through a nozzle to produce thrust. The accompanying figure shows a cutaway view of the solid-core nuclear thermal rocket, called NERVA (Nuclear Engine for Rocket Vehicle Applica-

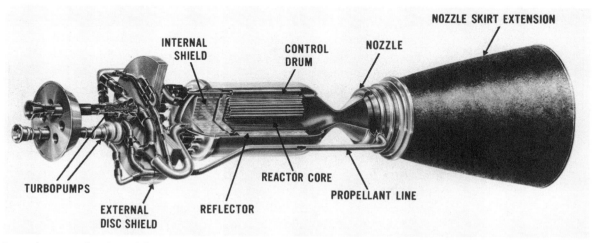

An explanatory drawing of the NERVA (Nuclear Engine for Rocket Vehicle Application) thermal nuclear rocket. From 1960 and 1973, this nuclear rocket engine was being developed under the supervision of the Space Nuclear Propulsion Office—a joint organization of NASA and the Atomic Energy Commission. Intended to support a human expedition to Mars in the 1980s, NERVA was cancelled in January 1973, prior to flight demonstration, due to changing national space objectives. *(NASA)*

tion). This system was developed by the United States between 1955 and 1973. However, because of changing national space program objectives, the development of this nuclear rocket engine ended just prior to flight demonstration.

Energy is released during the nuclear-fission process because the total mass of the fission products and neutrons after the reaction is less than the total mass of the original neutron and the heavy nucleus that absorbed it. From Albert Einstein's famous mass–energy equivalence relationship, $E = \Delta mc^2$, the energy released is equivalent to the tiny amount of mass that has disappeared (Δm) multiplied by the square of the speed of light (c).

The nuclear reactor is a device in which a fission chain reaction can be initiated, maintained, and controlled. Its essential component is the core, which contains the fissile fuel. Depending on the type of reactor, its purpose, and design, the reactor also can contain a moderator, a reflector, shielding, coolant, and control mechanisms. As shown in the figure, a nuclear rocket engine is much more than a heater of hydrogen. This type of rocket engine has five major segments. First, there is the nuclear reactor core, which serves as the heat source. Then, there are the turbopumps that pull liquid hydrogen from its tank and forces it through the reactor core. Next, there is the nozzle, which transforms the heated hydrogen gas to thrust. (The NERVA design also included a nozzle skirt extension to provide for more complete expansion of the exhausting hydrogen gas into the vacuum of outer space.) Fourth, the nuclear rocket requires a suitable structure to physically hold all the components together. Finally, the nuclear thermal rocket also uses a neutron reflector and control drum to regulate the fission chain reaction and various shields to protect the remainder of the spacecraft from the dangerous nuclear radiations released within the reactor core during operation and afterwards by decaying fission products.

Although some of the advanced nuclear rocket concepts currently remain on the distant technical horizon, the solid-core nuclear reactor rocket lies within a decade or so of flight-test demonstration and use on advanced space missions, such as a human-crewed expedition to Mars. The term *solid core* refers to the fact that the uranium-235 fuel in the reactor's core is kept in solid form (such as uranium carbide dispersed in graphite) during high-temperature operation. The heart of the NERVA engine was a reactor capable of releasing upward of 17,065 million Btu/h (5,000 megawatts) of thermal power, functioning for about one hour at a temperature around 2,500 K or 4,040°F (2,227°C) and generating 75,825 pounds-force (337,000 N) of thrust.

Achieving a critical mass with a solid core containing uranium-235 (as the nuclear fuel) is a matter of careful attention to neutron economics and reactor physics. Briefly summarized here, each fission (or split) of the uranium-235 nucleus releases, on average, 2.5 new neutrons. If the rate at which nuclear fissions occur in the core (and therefore the thermal power output) is to remain constant, exactly one of these 2.5 neutrons has to go on and cause another fission. Reactor criticality occurs once this point in neutron economics is achieved. At criticality, the neutron balance sheet leaves 1.5 neutrons per fission (on average) that can escape from the reactor altogether or else be absorbed in nonfission nuclear reactions.

To prevent too many neutrons from escaping, nuclear engineers place a material called a reflector around the reactor to bounce back or reflect some of the leaking neutrons back into the core. Graphite is a high-temperature material that reflects neutrons very well without absorbing too many of them. Engineers design the solid-core nuclear rocket so that there is enough uranium-235 dispersed through the graphite matrix to achieve a sustained chain reaction for the desired period of high-temperature (about 2,500 K or 4,040°F [2,227°C]) rocket engine performance.

The nuclear rocket's solid-core reactor is designed in such a way that all the fission-generated heat can be efficiently transferred to the hydrogen propellant, which is driven through the core by the turbopumps. If too much heat builds up in the uranium-graphite fuel matrix, the core temperature would quickly rise beyond the sublimation point. This condition would lead to deterioration or destruction of the core. To accomplish this enormous heat-removal task, engineers perforate the rocket reactor's fuel elements with precisely designed holes that run from the top to the bottom of the core and carry hydrogen through and past the hot fuel to the nozzle. These coolant holes or passages are coated with a thin layer of protective material, such as niobium carbide (NbC) or zirconium carbide (ZrC), to prevent chemical corrosion of the uranium-graphite fuel element by the hot hydrogen gas.

Cooling the core of a nuclear rocket engine is a challenging design task. If the engineers make the coolant passage holes too large, the size of the whole nuclear rocket engine becomes too large and unacceptable for use in space missions. On the other hand, if the engineers make the coolant passage holes too small, a great deal of friction and very high-pressure conditions occur as the hydrogen flows through them. Consequently, nuclear rocket engineers must make sure there are enough coolant passages with the optimum dimensions to efficiently remove all the heat from the reactor core.

To raise or lower the reactor's power level, the neutron economy or balance in the core must be changed. In a solid-core rocket reactor, like NERVA, control drums help accomplish this task. Control drums are cylinders covered on one side with a neutron poison (intense absorber) such as cadmium, boron, or hafnium. When all the neutron-absorbing faces of the control drums are turned inward, the neutron poison absorbs neutrons that would otherwise be reflected back into the reactor to cause new fission reactions. To start the nuclear rocket, motors slowly rotate the control drums, moving the neutron poison side away from the core and allowing more neutrons to participate in the chain reaction taking place in the core. Because successive neutron generations are only milliseconds apart, the neutron population (and therefore changes in reactor power level) can occur very quickly.

Nuclear rocket engine control is somewhat simplified by the fact that the hydrogen propellant itself has an impact on criticality. Specifically, when energetic fission neutrons collide with hydrogen nuclei, they slow down or moderate quickly to energies that more readily cause nuclear fission reactions in uranium-235 nuclei. This means that an increase in the flow of hydrogen through the rocket's reactor core is somewhat equivalent to slightly rotating the neutron poison side of the control drums slightly outward. On a human-crewed space mission that uses a nuclear rocket, the astronaut pilot would open the valve to the turbopump, effectively increasing both hydrogen flow through the core and the reactor power level. But this effect does not continue indefinitely. As the reactor power rises, so does the temperature of the hydrogen flowing through the core. Since the density of hydrogen decreases with temperature, there is now less hydrogen at a particular pressure and volume at any moment in the core—even though it is now being pumped through the core faster. Less hydrogen means that reactor criticality is affected less, and the power level begins to level off. Nuclear engineers say that the hydrogen has a negative temperature coefficient of reactivity.

Rocket engineers have several choices how to run the turbopumps that push hydrogen from the storage tank through the reactor core. The cold bleed cycle (as shown in the figure on page 132) diverts a small stream of hydrogen from the plenum above the reactor and circulates this initially cold hydrogen through heat transfer passages in the nozzle and reflector. The hydrogen gets heated as it flows through these passages and is then used to operate the main propellant feed pump before being exhausted to outer space. The hot bleed cycle extracts a small fraction (about 3 percent) of already heated hydrogen as it leaves the reactor core to run the turbopump. This particular approach was eventually selected for NERVA because it yielded the highest engine exhaust velocity.

NERVA development tests revealed that the cold bleed cycle made less efficient use of the hydrogen propellant. Specifically, with the cold bleed cycle more than 20 percent of the main hydrogen flow had to be extracted to drive the turbopump. This inefficient use of hydrogen would have greatly reduced overall performance of an operational nuclear rocket.

✧ Rover Program

The Rover Program was the name given to the overall United States nuclear rocket development program conducted from 1959 to 1973. At that time, space mission planners envisioned the use of the nuclear thermal rockets, placed in Earth orbit by giant Saturn V launch vehicles, to send human explorers to Mars sometime in the 1980s. The big selling point of the nuclear thermal rocket then (as now) is its inherently higher exhaust velocity, which is approximately double that of the best chemical rockets. This translates to quicker trips to Mars and/or more space vehicle mass arriving at Mars for the same starting mass placed in low Earth orbit at the beginning of the multiyear expedition. The use of a high–specific impulse nuclear propulsion system greatly reduces the overall transit time to Mars. This, in turn, alleviates strain on life support system reliability and also reduces the initial masses that must be delivered to low Earth orbit at the start of the expedition. Comparative studies have suggested that the ratio of the takeoff mass from Earth's surface to the final mass that achieves Earth escape velocity is a factor of about 15 for an all-chemical Mars rocket and about 3.2 for a Mars mission, which uses a solid-core nuclear-rocket upper stage.

The objective of the Rover Program/NERVA was to develop a flight-rated upper-stage nuclear rocket engine capable of providing 75,825 pounds-force (337,000 N) of thrust for a total operating period of about one hour—including shutdowns and restarts. The Rover Program began in 1955 when the U.S. Atomic Energy Commission (AEC) (now called the Department of Energy), the Los Alamos National Laboratory, and the U.S. Air Force began exploring the use of a nuclear rocket engine for application in an intercontinental ballistic missile (ICBM). However, on November 28, 1958, a chemically fueled Atlas ICBM successfully performed a full-range flight test and the U.S. Air Force soon withdrew its interest in any nuclear-rocket powered ICBM. That same year, NASA (the newly created civilian space agency) inherited the U.S. Air Force's investment in nuclear rockets and began examining the role of the nuclear rocket as the propulsion element in advanced, long-term space missions. The NERVA (Nuclear Engine for Rover Vehicle Application) portion of this overall Rover Program started two years later.

In 1960, NASA and the AEC created the Space Nuclear Propulsion Office to manage the overall Rover Program/NERVA effort. The Los Alamos National Laboratory served as the focal point for various prototype space reactor concepts and demonstrations, including Kiwi, Phoebus, Peewee, and the Nuclear Furnace (NF). Responsibility for the development of NERVA was assigned to an industrial team of engineers from the Aerojet General Corporation and Westinghouse Electric.

The Rover Program started with a group of Kiwi research reactors—named after the flightless bird found in New Zealand. Like their namesake, the Kiwi reactors were not intended to fly in space. Rather, scientists at Los Alamos National Laboratory used these pioneering research reactors to establish basic nuclear rocket–reactor technology and to demonstrate sound design approaches. For example, the Kiwi reactors were the first to operate with flowing liquid hydrogen and the first to demonstrate high temperature fuels. Kiwi reactors were also automatically controlled, using drums in the reflector. Although thermal-hydraulic vibration problems occurred in the early Kiwi reactors, scientists found a solution for these problems. The Kiwi program culminated in the Kiwi-B4E reactor, which operated at over 1,890 K or 2,942°F (1,617°C) for 11.3 minutes and at 2,005 K or 3,149°F (1,732°C) for 95 seconds at a power level of 3,200 million Btu/h (937 megawatts [MW]).

The Kiwi reactors led to the NRX series of development reactors. Their goal was to demonstrate a specific impulse of 760 seconds (7,450 m/s in SI units) for a period of 60 minutes. These program objectives were exceeded in the NRX-A6 test, which ran for 62 minutes at a temperature of 2,220 K or 3,536°F (1,947°C) and a power level of 3,754 million Btu/h (1,100 MW).

Los Alamos scientists also developed another series of rocket-related research reactors, called Phoebus. As they built the family of Phoebus reactors, scientists wanted to increase the specific impulse to 825 seconds (8,085 m/s in SI units) and increase the power level to between 13,650 million Btu/h (4,000 MW) and 17,065 million Btu/h (5,000 MW). These technical milestones were well demonstrated in tests of the Phoebus-1B and Phoebus-2A systems. For example, the Phoebus 2A operated for over 12 minutes above 13,650 million Btu/h (4,000 MW)—reaching a peak power level of 14,000 million Btu/h (4,100 MW).

The Peewee and Nuclear Furnace (NF) reactor series demonstrated the performance of longer-lived, high-temperature nuclear rocket engine fuel elements. Peewee-1 operated at 2,555 K or 4,139°F (2,282°C) for 40 minutes, generating a total power of 1,754 million Btu/h (514 MW). NF-1 ran at 2,450 K or 3,950°F (2,177°C) for 109 minutes at a power level of 184 million Btu/h (54 MW).

In 1965, the nuclear rocket reactor test program introduced a new class of reactors aimed at generating power in the 13,650 million Btu/h (4,000 MW) and 17,065 million Btu/h (5,000 MW) range. Scientists and engineers used experimental engines to determine how a nuclear rocket would behave during startup, full power operation, and shutdown. They also used this test program to evaluate control concepts and to qualify engine test stand operations in a downward firing mode—that is, with the nozzle at the bottom of the test engine. These objectives were met or exceeded in the NRX/EST and XE test programs. For example, XE—a prototype flight engine system—was tested in a simulated space environment and performed some 28 starts and restarts. In this engine, nonnuclear flight components were tested along with a flight-type reactor.

In 1969, engineers tested the first prototype nuclear rocket engine in a downward firing mode at the Nuclear Rocket Development Station in Nevada. Their test configuration included a simulated space environment. However, that same year, NASA suspended production of the Saturn V launch vehicle. Since the Saturn V vehicle was the prime launch vehicle for the nuclear rocket, this cancellation heralded the national decision to abandon the human exploration of Mars as a direct follow-up to the highly successful Apollo lunar landings.

There was a final test of the Nuclear Furnace (NF) system in 1972, during which the system demonstrated operation at power densities of approximately 439 million Btu/h-ft^3 (4,500 MW/m^3) and temperatures in excess of 2,500 K or 4,040°F (2,227°C) for about 109 minutes. Then, despite all its technical achievements, the United States government terminated the nuclear rocket program in January 1973. Rover/NERVA was a casualty of budget cuts and changing national space mission emphasis.

✧ Nuclear Rocket Development Station

From 1959 to 1973, the Nuclear Rocket Development Station (NRDS) at the Nevada Test Site (NTS) in southern Nevada housed the major test facilities for the U.S. nuclear rocket program. The NRDS was situated on a desolate flat basin called Jackass Flats after some of the indigenous wildlife. Both nuclear reactors and complete nuclear rocket engine assemblies were tested (in place) at the station, which had three major test areas: Test cells A and C and Engine Test Stand Number One (ETS-1). The reactor test facilities were designed to test a reactor in an upward-firing position, while the engine test facility could test prototype nuclear rocket engines in a downward-firing mode.

The three test cell areas were connected by road and a designated railroad line, called the "Jackass & Western Railroad." Scientists and engineers humorously referred to the railroad as the world's shortest and slowest. A research reactor or prototype rocket engine would be assembled in one of the Maintenance Assembly and Disassembly (MAD) buildings at the station and then carefully transported by the Jackass & Western Railroad to one of the test locations. After a test, while the research reactor or nuclear rocket engine was still very radioactive (because of the buildup of fission products in the core), a heavily shielded railroad engine would tow it back to the proper MAD building for disassembly. The MAD buildings were basically immense hot cells where engineers and scientists, protected by thick concrete and stainless steel radiation shields, could perform robot-assisted disassembly of the radioactive reactor cores and prototype nuclear rocket engines.

During a nuclear rocket test at the Nuclear Rocket Development Station, the surrounding desert basin literally became an inferno, as very hot hydrogen gas spontaneously ignited upon contact with the surrounding air and reacted with atmospheric oxygen to form water. Unlike the highly visible, radiant "pillar of fire" that accompanies the operation of a typical chemical rocket engine, a nuclear rocket engine's exhaust is essentially invisible—during ground-based testing—except perhaps for any

A newly assembled Phoebus-2A nuclear rocket research reactor being transported by the Jackass & Western Railroad to its test cell at the Nuclear Rocket Development Station in Nevada, ca. 1967 *(U.S. Department of Energy/ Los Alamos National Laboratory)*

The Kiwi–A nuclear rocket research reactor being tested at the Nuclear Rocket Development Station in 1958. During the test, the surrounding desert basin became a literal inferno as intensely hot hydrogen gas exhausted the Kiwi's nozzle and came in contact with the surrounding air. The hot hydrogen immediately reacted with (burned) atmospheric oxygen to form water, as indicated by the steam plume in the center of the picture. (*U.S. Department of Energy/Los Alamos National Laboratory*)

incandescent impurities that end up in the hot hydrogen exhaust and for thermal aberrations produced in the surrounding air as the intensely hot hydrogen exits and forms clouds of steam.

✧ Project Pluto

In January 1957, while searching for intercontinental ballistic missile propulsion systems, the U.S. Air Force and the Atomic Energy Commission selected the Lawrence Radiation Laboratory (forerunner of the current Lawrence Livermore National Laboratory) to study the feasibility of applying heat from a compact nuclear reactor to operate a ramjet engine capable of hurling a guided missile across continental distances. This line of research took place under the code name Project Pluto. Work on the nuclear propulsion project soon moved from Livermore, California to special test facilities at Nuclear Rocket Development Station in Nevada.

The Project Pluto test complex consisted of six miles (10 km) of roads, a critical assembly building, a control building, various assembly and shop buildings, a railroad, and utilities. The unusual project also required the construction of 25 miles (40 km) of oil well casing—in order to store the half-million kilograms of pressurized air needed to simulate ramjet flight conditions.

The principle behind the nuclear ramjet is relatively simple: Air is drawn in at the front of the vehicle under ram pressure (that is, under great force due to the vehicle's high-speed motion), heated to extremely high temperatures as it passes through a compact, mobile nuclear reactor, and then expanded through a nozzle and exhausted out the back to provide thrust.

The Project Pluto reactor had to be small and compact enough to fly, but durable enough to survive a 6,800-mile (11,000-km) flight to a potential target. The success of this project depended upon a series of technological advances in metallurgy and materials science. In addition, pneumatic motors necessary to control the reactor in flight had to operate while red-hot and in the presence of intense radioactivity. The need to maintain supersonic speed at low altitude and in all kinds of weather conditions meant the reactor, code-named Tory, had to survive temperatures on the order of 1,700 K or 2,600°F (1,427°C) and operate under conditions that would melt the metals used in most jet and rocket engines.

On May 14, 1961, the world's first nuclear ramjet engine, called Tory-IIA, was mounted on a railroad car at the Nevada Test Site and roared to life for just a few seconds. Despite other successful engineering tests, the U.S. Air Force cancelled Project Pluto on July 1, 1964. The Department of Defense had successfully developed and deployed chemical rockets that could adequately perform the ICBM mission. So, a nuclear ramjet speeding at low-altitude through Earth's atmosphere simply came with too much political, environmental, and technical baggage to merit further study or fiscal support.

✧ Nuclear Rockets in the 21st Century

In 2003, NASA established Project Prometheus to develop technology and conduct advanced studies involving the use of nuclear power and propulsion systems in support of space exploration this century. There is also renewed White House interest (as announced by President George W. Bush in January 2004) in sending human beings back to the Moon and then on to Mars.

Over the next two decades, as these initiatives exert influence on the overall shape and structure of the American civilian space program, the nuclear thermal rocket should once more become the interplanetary propulsion system of choice for the first human expedition to Mars. Nuclear propulsion can shorten interplanetary trip times and reduce the expedition mass launched from Earth. As the primary propulsion system for interplanetary transfer, the space vehicle's nuclear thermal rocket would remain inactive until it was time for the explorers to depart from Earth orbit. Selective nuclear rocket engine firings would send their spacecraft on a trajectory to Mars, insert the expedition spacecraft in orbit around Mars, and then return the crew back to the vicinity of Earth. The role of the nuclear thermal rocket during this historic mission would end when the crew fired it for the final time to insert their spacecraft into orbit around Earth. Following this final orbit insertion maneuver, the crew would travel to the surface of Earth using any one of several candidate reentry/landing systems.

As a permanent lunar base grows, routine 24-hour flights to the Moon might use detachable crew modules that ride atop nuclear thermal

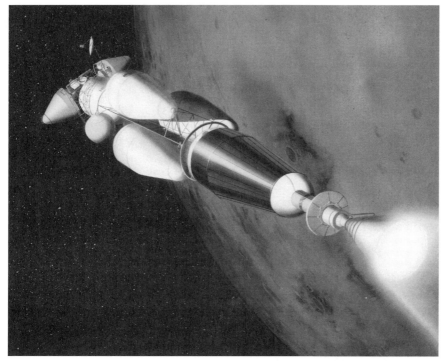

This artist's concept shows a nuclear thermal rocket performing an orbit insertion firing, as the spacecraft carrying the first human expedition to Mars arrives in the vicinity of the Red Planet. *(Artwork courtesy of NASA; artist: Pat Rawlings, 1995)*

In this artist's concept on the way to the Jovian system, a nuclear thermal rocket–powered interplanetary transfer vehicle refuels in orbit around Mars near the Red Planet's moon Phobos. *(Artwork courtesy of NASA; artist: Pat Rawlings, 1996)*

rocket orbital transfer vehicles. By transferring the crew module from one propulsion element to the next, the passengers could complete their trip from low Earth orbit to the surface of the Moon without ever leaving the module.

As human explorers push beyond Mars into the main asteroid belt and to the vicinity of certain intriguing Jovian moons, the nuclear thermal rocket would represent an enabling propulsion technology. As depicted in the figure above, the planet Mars could become a convenient refueling station. Permanent human settlements on Mars and its natural moons (Phobos and Deimos) might serve as "frontier towns," capable of outfitting future expeditions that push human presence to the very edges of the solar system. Technical descendents of the Rover Program/NERVA would become a fleet of powerful, reliable, and long-range nuclear thermal rocket–powered interplanetary transfer vehicles—transporting human beings and cargo to a variety of destinations throughout the solar system.

Electric Propulsion for Deep-Space Missions

This chapter presents the basic physical principles that govern the operation of the electric rocket. Contemporary and planned future use of electric propulsion for deep space missions is also described.

The electric rocket engine is a device that converts electric power into a forward-directed force or thrust by accelerating an ionized propellant (such as mercury, cesium, argon, or xenon) to a very high exhaust velocity. The concept for an electric rocket is not new. In 1906, the American rocket scientist, Robert H. Goddard, suggested that the exhaust velocity limit encountered with chemical rocket propellants might be overcome, if electrically charged particles could be used as a rocket's reaction mass. Technical historians often regard his technical suggestion as the birth of the electric propulsion concept.

The basic electric propulsion system consists of three main components: some type of electric thruster that accelerates the ionized propellant, a suitable propellant that can be ionized and accelerated, and a source of electric power. The acceleration of electrically charged particles requires a large quantity of electric power.

The needed power source could be self-contained, such as a space nuclear reactor, or it might involve the use of solar energy by means of photovoltaic or solar thermal conversion techniques. Electric propulsion systems using a nuclear reactor power supply are called nuclear electric propulsion (NEP) systems, while those using a solar energy power supply are called solar electric propulsion (SEP) systems. Within the orbit of Mars, both NEP and SEP systems can be considered, but well beyond Mars and especially for deep space missions to the edges of the solar system only nuclear electric propulsion systems appear practical within

21st century technology horizons. This is due to the fact that the amount of solar energy available for collection and conversion falls off according to the inverse square law—that is, as one over the distance from the Sun squared [$1/(\text{distance})^2$].

✧ Fundamentals of Electric Propulsion

There are three general types of electric rocket engine: electrothermal, electromagnetic, and electrostatic. In the basic electrothermal rocket, electric power is used to heat the propellant (such as ammonia) to a high temperature. The heated propellant is then expanded through a nozzle to produce thrust. Propellant heating may be accomplished by flowing the propellant gas through an electric arc (this type of electric engine is called an arc jet engine) or by flowing the propellant gas over surfaces heated with electricity.

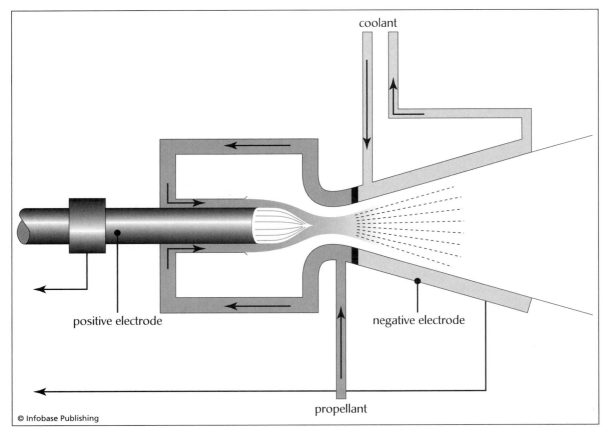

© Infobase Publishing

Basic components of an electrothermal rocket

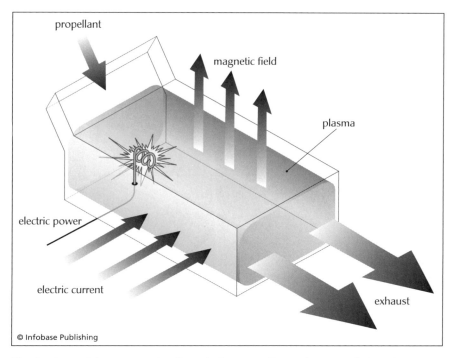

propellant

magnetic field

plasma

electric power

electric current

exhaust

© Infobase Publishing

The fundamental components of an electromagnetic or plasma rocket engine

Although the arc jet engine can achieve exhaust velocities higher than those of chemical rockets, the dissociation of propellant gas molecules creates an upper limit on how much energy can be added to the propellant. In addition, other factors, such as erosion caused by the electric arc itself and material failure at high temperatures establish further limits on the arc jet engine. Because of these limitations, arc jet engines are more suitable for a role in orbital transfer vehicle propulsion and large spacecraft station keeping than as the electric propulsion system for deep space exploration missions.

The second major type of electric rocket engine is the electromagnetic engine or plasma rocket engine. In this type of engine, the propellant gas is ionized to form a plasma, which is then accelerated rearward by the action of electric and magnetic fields. The magnetoplasmadynamic (MPD) engine can operate in either a steady state or a pulse mode. A high-power (approximately 3.4 million Btu/h [1 megawatt-electric]) steady-state MPD, using either argon or hydrogen as its propellant, is an attractive option for an electric propulsion orbital transfer vehicle (OTV).

The third major type of electric rocket engine is the electrostatic rocket engine or ion rocket engine. As in the plasma rocket engine,

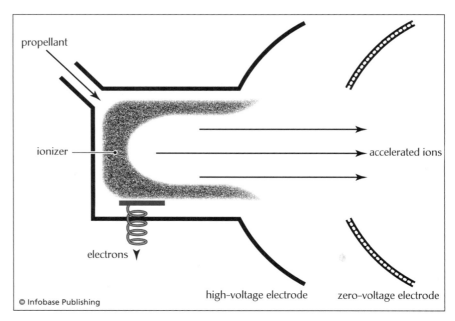

The fundamental components of an electrostatic or ion rocket engine

propellant atoms (i.e., cesium, mercury, argon, or xenon) are ionized by removing an electron from each atom. In the electrostatic engine, however, the electrons are removed entirely from the ionization region at the same rate as ions are accelerated rearward. The propellant ions are accelerated by an imposed electric field to a very high exhaust velocity. The electrons removed in the ionizer from the propellant atoms are also ejected from the spacecraft, usually by being injected into the ion exhaust beam. This helps neutralize the accumulated positive electric charge in the exhaust beam and maintains the ionizer in the electrostatic rocket at a high voltage potential.

In 1970, NASA successfully tested two 5.9-inch- (15-cm-) diameter mercury-propellant ion thrusters in space onboard the *SERT 2* (*Space Electric Rocket Test–2*) spacecraft, which was placed in a 620-mile (1,000-km) altitude, sun-synchronous polar orbit. Each of these ion thrusters provided a maximum thrust of about 0.006 pound-force (26.7 millinewton [mN]). (A millinewton is a very tiny force that equals one thousandth of a newton.) The extended operation of the two thrusters demonstrated long-term ion thruster performance in the near-Earth orbital environment, but also introduced aerospace engineers to the problem of "sputtering." Sputtering involves the buildup of molecular metal contaminants on the host spacecraft. For example, mercury accumulated on the solar arrays of the *SERT 2* spacecraft, as a result of ion thruster exhaust plume contamination. Thruster experimentation with the *SERT 2* spacecraft ended in 1981, when the mercury propellant supply became exhausted.

NASA's electric propulsion technology program next focused on the J-series 11.8-inch- (30-cm-) diameter ion thrusters, using mercury as the propellant. This type of electric rocket represented reasonably mature technology with certain restrictions. Because of the potential for pollution and contamination (i.e., sputtering phenomenon), aerospace engineers now regard mercury as an unacceptable propellant for the heavy OTV

This image of a xenon ion engine was photographed through the viewing port that peeks into the vacuum chamber, where engineers at NASA's Jet Propulsion Laboratory operated the test electric rocket in a cyclic manner (that is, two days on, one hour off, then restart) for a total of 8,000 hours in 1999. The faint blue glow results from the charged atoms of xenon emitted by the ion engine. *(NASA/Jet Propulsion Laboratory)*

traffic that should operate from low Earth orbit (LEO) to various destinations in cislunar space later this century.

To overcome these limitations, propulsion engineers turned their attention to ion thrusters using argon or xenon as the propellant. Because of the potential for providing very high exhaust velocities (typically 32,800 ft/s [10,000 m/s]) and very high efficiency, xenon-fueled ion propulsion systems—like the one shown in the figure on page 149—appear well suited for cislunar OTV missions, as well as interplanetary and deep-space exploration missions.

Unlike the fireworks that accompany most solid or liquid-fueled chemical rockets, the ion engine emits only an eerie blue glow—seen as electrically charged (ionized) atoms of xenon are expelled from the engine. Xenon is the same gas found in photo flash tubes and many high-intensity lighthouse bulbs. Another very important distinction between the chemical rocket and the electric rocket is the level and duration of thrust produced by each type of reaction engine. The chemical rocket generally delivers a large amount of thrust (typically thousands of pounds-force [kilonewtons]) in a relatively short period of time (seconds to minutes). In contrast, an ion engine operates continuously—often for weeks and months, delivering an almost imperceptible level of thrust—typically on the order of a few thousandths of a pound-force (millinewtons). The engine shown in the figure on page 149, for example, delivers a minuscule thrust that is roughly equivalent to the pressure (force per unit area) exerted by a sheet of paper held in the palm of a person's hand. Since the ion engine is very slow to pick up speed, it cannot lift objects into space from the surface of a planet. But when operated in space for extended periods, the ion engine typically delivers 10 times as much thrust per pound-mass (kilogram) of propellant as more traditional chemical rockets over the long haul.

✧ Deep Space One

NASA's *Deep Space One* (DS-1) technology demonstration space probe was powered by two solar panel wings and an electric propulsion system. This particular ion engine represented the first nonchemical propulsion system used by NASA as the primary means of propelling a spacecraft. The robot spacecraft's 11.8-inch (30-cm) diameter xenon ion engine used approximately 6,825 Btu/h (2,000 watts) of solar cell–generated electric power to ionize the xenon gas and then accelerate these ions to about 103,320 ft/s (31,500 m/s). By ejecting these high-speed ions, the electric rocket produced some 0.02 pounds-force (0.09 N) of thrust. The propellant tank contained an initial supply of 179 pounds (81.5 kg) of xenon gas. This demonstration engine was also capable of providing about 0.0045 pounds-force (0.02 N) of thrust when functioning at the minimum operational electrical power level of 1,707 Btu/h (500 watts [W]).

Deep Space One was launched from Pad 17-A at the Cape Canaveral on October 24, 1998, by a Delta II chemical rocket. Once in space, the spacecraft's ion engine began to operate and allowed DS-1 to perform a flyby of the near-Earth asteroid 9969 Braille on July 29, 1999, at a distance of about 16 miles (26 km) and a relative velocity of approximately 9.6 miles per second (15.5 km/s). By the end of 1999, the ion engine had used approximately 48 pounds (22 kg) of xenon to impart a total velocity change (called delta V) of 4,264 feet per second (1,300 m/s) to the spacecraft.

The original plan was to fly by the dormant comet Wilson-Harrington in January 2001 and then past the comet Borrelly in September 2001. But the spacecraft's star tracker failed on November 11, 1999, so mission planners drew upon techniques developed to operate the spacecraft without the star tracker and came up with a new extended mission to fly by Comet Borrelly. As a result of these innovative actions, on September 22, 2001, *Deep Space One* entered the coma of Comet Borrelly and successfully made its closest approach (a distance of about 1,366 miles [2,200 km]) to the nucleus at about 22:30 UT (universal time). At the time of the comet encounter, DS-1 was traveling at 10.25 miles per second (16.5 km/s) relative to the nucleus. Following this encounter, NASA mission controllers commanded the spacecraft to shut down its ion engines on December 18, 2001. Their action ended the *Deep Space One* mission. All the space technologies flown on board DS-1, including the 11.8-inch (30-cm) diameter xenon ion engine, were successfully tested during the primary mission.

✧ Nuclear Electric Propulsion Systems

Space visionaries, starting with Robert Goddard, have recognized the special role electric propulsion can play in the conquest of space—namely, high-performance missions starting in a low gravity field (such as Earth orbit or lunar orbit) and the vacuum of free space. In comparison to high-thrust, short-duration-burn chemical engines, electric propulsion systems are inherently low-thrust, high-specific-impulse rocket engines with fuel efficiencies two to 10 times greater than the propulsion efficiencies achieved using chemical propellants. Electric rockets work continuously for long periods, smoothly changing a spacecraft's trajectory. For missions to the outer solar system, the continuous acceleration provided by an electric propulsion thruster can yield shorter trip times and/or deliver higher-mass scientific payloads than those delivered by chemical rockets.

This artist's rendering depicts NASA's proposed nuclear fission reactor–powered, ion engine–propelled spacecraft entering the Jovian system, ca. 2015. The Jupiter Icy Moons Orbiter (JIMO) mission would perform detailed scientific studies of Callisto, Ganymede, and Europa (in that order), searching for liquid water oceans beneath their surfaces. Europa is of special interest to the scientific community because its suspected ocean of liquid water could contain alien life-forms. *(NASA/Jet Propulsion Laboratory)*

NASA is now developing plans for an ambitious mission to orbit three planet-sized moons of Jupiter (Callisto, Ganymede, and Europa), which may harbor vast oceans beneath their icy surfaces. The Jupiter Icy Moons Orbiter (JIMO) mission would raise NASA's capability for space exploration to a revolutionary new level by pioneering the use of electric propulsion powered by a nuclear fission reactor.

The *JIMO* spacecraft would pioneer the use of electric propulsion powered by a nuclear fission reactor. Contemporary electric rocket technology—successfully tested on the NASA's *Deep Space One (DS-1)* spacecraft—now allows NASA planners to design *JIMO* spacecraft to orbit three different moons during a single mission. Current spacecraft, like *Cassini,* have enough onboard propulsive thrust capability (upon arrival at a target planet) to orbit that single planet and then use various orbits to fly by any moons or other objects of interest, such as ring systems. In contrast, *JIMO*'s nuclear electric propulsion system would have the necessary long-term thrust capability to gently maneuver through the Jovian system and allow the scientific spacecraft to successfully orbit and carefully investigate each of the three icy moons of interest.

In one conceptual design for the *Jupiter Icy Moons Orbiter* spacecraft electric power is supplied by a compact nuclear fission reactor. A 65.6-foot (20-m) boom provides mechanical support for the large array of heat radiator panels, which are needed to reject waste thermal energy to outer space. This large boom also separates the science payload from the vicinity of the nuclear reactor. Finally, the *JIMO* spacecraft has two thruster pods, each of which would contain a bank of 10 11.8-inch- (30-cm-) diameter xenon ion engines. The compact nuclear fission reactor would generate unprecedented quantities of electricity—significantly improving scientific measurements, mission design options, and telecommunications. With sufficient spacecraft electrical power, scientists can expand the capabilities of the JIMO mission to include a power radar that can penetrate deep into the icy surfaces of the target Jovian moons and also perform extensive, high-resolution mapping of surface features. The *JIMO* electric propulsion system will deliver up to 100 times more thrusting capability than a non-nuclear electrically propelled spacecraft of comparable (overall) mission mass.

Nuclear electric propulsion technology makes it possible for scientists to design a single mission for orbiting three very interesting moons of Jupiter, one after another. This type of electric rocket also opens up the rest of the outer solar system to detailed exploration in later missions. Unlike the solar electric propulsion (SEP) system, the nuclear electric propulsion (NEP) system can operate anywhere in the solar system and even beyond, since its performance is independent of its position relative to the Sun. A NEP system can provide shorter trip times and greater

This artist's rendering shows a football field–sized nuclear electric–propelled vehicle firing banks of ion engines in order to circularize its orbit around Mars. This interplanetary transfer vehicle would be assembled in Earth orbit, take about six and a half months to reach Mars, and be capable of hauling 286,000 pounds (130,000 kg) of payload in support of permanent human settlements on the surface of the Red Planet (circa 2050). *(Artwork courtesy of NASA; artist: Pat Rawlings, 1992)*

payload capacity than any of the advanced chemical propulsion technologies available in the next two decades or so for detailed exploration of the outer planets—especially Saturn, Uranus, Neptune, Pluto, and their respective moons.

Larger, more advanced-technology, nuclear electric propulsion systems could become the main space-based transport vehicles for both human explorers and their equipment as permanent surface bases are established on Mars in the mid-21st century. The figure above is an artist's rendering of a large nuclear electric propelled vehicle (about the size of a football field), firing banks of ion engine thrusters in order to circularize its orbit around Mars. Assembled in Earth orbit, the transfer vehicle, with its 34.1 million Btu/h (10 megawatt) nuclear power plant, could transport a payload with a mass of 286,000 pounds (130,000 kg) to Mars in six and one half months and could repeat the interplanetary circuit every 52 months.

Advanced Propulsion Systems for the 21st Century

This chapter describes some of the relatively near-term propulsion system developments that could greatly improve either access to space from Earth's surface or else in space transportation from one orbital location to another. The technical spectrum of future space launch vehicles ranges from stepwise improvements in current chemical rocket systems—such as are now taking place within the United States Air Force's evolved expendable launch vehicles program—to major transformations, such as would occur with the arrival of completely reusable launch vehicles and magnetic levitation-assisted (maglev) rocket vehicles. As discussed in chapter 6, NASA plans to replace the shuttle by 2010 with two new shuttle-derived vehicles: a crew-carrying rocket system and a heavy-lift, cargo-carrying rocket system.

Future space launch options also include more radical concepts, such as the gun-type launch system. In-space transportation will enjoy significant improvements with advances in nuclear and electric propulsion systems (discussed in chapters 9 and 10), as well as with such interesting approaches as laser propulsion systems, the solar sail, the plasma sail, and aerocapture technologies.

✧ Evolved Expendable Launch Vehicle

The evolved expendable launch vehicle (EELV) is the space-lift modernization program of the United States Air Force. EELV is reducing the cost of launching payloads into space by at least 25 percent over current Delta, Atlas, and Titan launch systems. Part of these savings result from the government now procuring commercial launch services and turning over responsibility for operations and maintenance of the launch complexes to the contractors. This new space-lift strategy has reduced the government's traditional involvement in launch processing, while saving a projected $6 billion in launch costs between the years 2002 and 2020. In addition, EELV improves space launch operability and standardization.

The mission statement for the EELV program is, "Partner with industry to develop a national launch capability that satisfies both government and commercial payload requirements and reduces the cost of space launch by at least 25 percent." The EELV program's two primary objectives are: first, to increase the U.S. space launch industry's competitiveness in the international commercial launch services market, and second, to implement acquisition reform initiatives resulting in reduced government resources necessary to manage system development, reduced development cycle time, and deployment of commercial launch services.

There are two primary launch contractors in the EELV program: the Lockheed Martin Company and Boeing Company. The Atlas V launch vehicle results from the culmination of Lockheed Martin's desire to employ the best practices from both the Atlas and Titan rocket programs into an evolved and highly competitive commercial and government launch system for this century. The Atlas V launch vehicle builds on the design innovations that will be demonstrated on Atlas III and incorporates a structurally stable booster propellant tank, enhanced payload fairing options, and optional strap-on solid rocket boosters.

The Boeing Company's Delta IV family of expendable launch vehicles blends new and mature technology to lift virtually any size medium or heavy payload into space. This family of rockets consists of five vehicles based on a common booster core (CBC) first stage. Delta IV second stages are derived from the Delta III second stage, using the same RL10B-2 engine, but with two sizes of expanded fuel and oxidizer tanks, depending on the model. In designing the five Delta IV configurations, the Boeing Company conducted extensive discussions with government and commercial customers concerning their present and future launch requirements. Proven technical features and processes were carried over from earlier Delta vehicles to the Delta IV family of space lift vehicles.

The first three EELV missions were all successfully launched from Cape Canaveral Air Station, Florida. The inaugural launch of Lockheed-Martin's Atlas V Medium Lift Vehicle took place on August 21, 2002 and placed a commercial Eutelsat payload to orbit. Three months later, on November 20, 2002, a Boeing Delta IV rocket successfully placed another commercial Eutelsat payload into orbit around Earth. Eutelsat spacecraft are commercial communications satellites that operate in geostationary orbit. The Boeing Company also accomplished the first government launch of an EELV—the Delta IV rocket that carried the *DSCS A3* military communications satellite into orbit on March 10, 2003. The first launch of the Heavy Lift Vehicle (HLV) version of the Delta IV took place from Cape Canaveral on December 21, 2004. Sponsors view the overall mission as a success, although the rocket's main-stage propulsion system experienced some problems and prematurely shutdown.

✦ Reusable Launch Vehicle

The reusable launch vehicle (RLV) is an aerospace vehicle that incorporates functional designs and fully reusable components to achieve airline-type spaceflight operations simple, fully reusable designs for airline-type operations using advanced technology and innovative operational techniques.

With the exception of the U.S. Space Transportation System—that is, NASA's space shuttle (see chapter 5)—all modern rockets currently used to lift payloads into space are expendable launch vehicles (ELVs). This means that all the rocket vehicle's flight components, such as engines, fuel tanks, and support structures, are discarded after just one use. On the one hand, the practice of using "throwaway" launch vehicle makes access to space quite expensive. On the other hand, the practice also accommodates incremental improvements in a particular family of expendable launch vehicles. Current American launch costs, using both ELVs and the space shuttle, range between $2,500 per pound ($5,000 per kg) and $10,000 per pound ($20,000 per kg) of mass delivered into low Earth orbit—with space shuttle–delivered payloads dominating the upper end of the cost spectrum. Of course, the exact expense of placing an object in space depends on such factors as the desired orbital altitude and inclination, the characteristics of the launch vehicle, and the need (if any) for upper-stage vehicles.

NASA's space shuttle is only a partially reusable aerospace vehicle, which requires extensive refurbishment between launches, and is extremely costly to operate. The American aerospace industry first seriously turned its attention to a reusable space transportation system between 1968 and 1971. Unfortunately, design compromises and budget reductions during the space shuttle's development in the 1970s eroded many of the reusability features and launch-cost savings originally envisioned for this system.

A 1995 partnership with the U.S. Air Force and private industry enabled NASA to create the Reusable Launch Vehicle Technology Program. The main objective of that program was to develop and demonstrate new technologies for the next generation of fully reusable space transportation systems that can radically reduce the cost of accessing space—perhaps to as low as $50 per pound ($100 per kilogram) delivered to low Earth orbit (LEO). Three candidate RLV designs were considered: the lifting body configuration, the vertical landing configuration, and the winged body configuration. The goal of this effort (now canceled) was to demonstrate the technology for an all-rocket-powered single-stage-to-orbit (SSTO) RLV.

Each candidate SSTO RLV configuration presents unique technical obstacles, but also offers some distinct advantages for reducing the cost of accessing space. The conical configuration, for example, provides the advantage of a low-mass airframe that does not require enormous wings and offers a simple aerodynamic shape. But this candidate reusable rocket vehicle's engines must be restarted after reentry into Earth's atmosphere to permit a vertical landing. In addition, the conical configuration has a limited payload volume. The winged body SSTO configuration offers the advantage of simplified fuel tanks and easy maneuverability during reentry. Its disadvantages include a high landing speed and a limited payload volume. The advantages of the lifting body SSTO configuration include

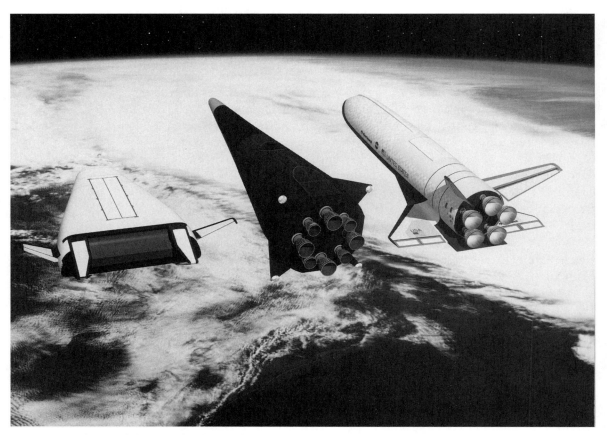

An artist's rendering of the three fundamental reusable launch vehicle (RLV) single-stage-to-orbit (SSTO) concepts (from left to right): the lifting body configuration, the vertical landing configuration, and the winged body configuration *(NASA/Marshall Space Flight Center)*

a low-mass design, a low landing speed, and low reentry temperatures. One of the main disadvantages of this configuration is its complicated airframe.

NASA, in partnership with private industry and the U.S. Air Force, pursued the winged body configuration in a prototype RLV technology demonstration program called the X-33. NASA's X-33 program attempted to simultaneously demonstrate several major advancements in space launch vehicle technology that would increase safety and reliability while lowering the cost of placing a payload into low earth orbit by an order of magnitude or so—that is, from about $10,000 per pound ($20,000 per kg) to perhaps as little as $1,000 per pound ($2,000 per kg). The X-33 vehicle was a half-scale prototype of the conceptual single-stage-to-orbit (SSTO) RLV, called VentureStar, being pursued by the Lockheed-Martin Company as a commercial development. However, after a series of disappointing technical setbacks and serious cost overruns, NASA abruptly cancelled the X-33 program in March 2001.

There is a contemporary NASA focus on RLV technology development within a program called the Space Launch Initiative (SLI). This NASA program involves work with the U.S. aerospace industry to design a privately operated second generation RLV that will reduce "loss of crew" and mission risks and is more cost-effective compared to today's space shuttle. Aerospace experts consider it possible to build an RLV reasonably early in this century. Yet, achieving the performance levels to reach low Earth orbit with a useful payload will require a host of engineering advances that improve fuel efficiency and lower vehicle mass without compromising structural integrity. Completely reusable launch vehicles are much more difficult to construct than expendable (throwaway) rockets, because all of an RLV's components must be capable of resisting deterioration and surviving numerous launches and reentries without requiring costly and time-consuming refurbishment between flights. The RLV is perceived as the Holy Grail of chemical rocketry, because, if developed, it would provide routine, low cost (possibly on the order of $50 per pound [$100 per kg] delivered to low Earth orbit) access to space—opening up the "space frontier" to many new users and additional applications.

The figure on page 162 is an artist's concept of a maglev (magnetic levitation) assisted third generation of reusable launch vehicle. Some of NASA's long-range planners think an efficient way into space is to give the future RLV an electromagnetic "running start." A long track would use magnetic fields to levitate and accelerate the reusable launch vehicle to about 0.16 mile per second (0.25 km/s) before the vehicle's liquid-propellant (hydrogen-oxygen) rocket engines ignited. Another variation being considered for the third-generation RLV is a magnetic levitation (or maglev)–assisted launch system in which the magnetically accelerated flight vehicle first

This artist's concept shows a magnetic levitation (maglev)–assisted third–generation reusable launch vehicle firing up its engines after being accelerated to a near-supersonic speed of 0.25 km/s–circa 2025. *(NASA/Marshall Space Flight Center)*

fires an air-breathing ramjet. As the aerospace vehicle increases in speed and climbs higher in altitude, the ramjet engine shuts down and a liquid hydrogen-liquid oxygen rocket engine ignites, providing the final thrust into low Earth orbit.

Magnetic levitation (maglev) employs well-understood principles of magnetism to levitate a vehicle so that it does not touch anything during acceleration. Magnets have two poles: one north and the other south. Opposite poles attract each other, while similar poles repel each other. By using the phenomenon of similar pole repulsion, engineers can levitate a vehicle and make it accelerate more smoothly, since contact friction is greatly reduced. High-speed maglev trains are an example. In Germany, a maglev train has smoothly traveled at 270 miles per hour (435 km/h). Encouraged by these developments, aerospace engineers are now examining the maglev-assisted reusable launch vehicle as a way of significantly reducing the cost of accessing space in a decade or so.

✧ Gun–Launch–to–Space

Gun-launch-to-space (GLTS) is an advanced space-access concept that involves the use of a long and powerful electromagnetic launcher to hurl small satellites and payloads into orbit. An electromagnetic launcher is a device that can accelerate an object to high velocities using the electromotive force produced by a large current in a transverse magnetic field.

One recent version of the GLTS concept suggests the development of a hypervelocity coil-gun launcher to place 220-pound (100-kg) payloads into Earth orbit at altitudes ranging from 125 miles (200 km) to 310 miles (500 km). This coil-gun launcher would accelerate the specially designed (and acceleration-hardened) payload package to an initial velocity of about 3.7 miles per second (6 km/s) through a long, evacuated tube. The payload package would consist of the payload itself (e.g., a satellite or bulk cargo), a solid-propellant orbital insertion rocket, a guidance system, and an aeroshell. The specially designed payload package would penetrate easily through the atmosphere and be protected from atmospheric heating by its aeroshell. Once out of the sensible atmosphere, the aeroshell would be discarded and the solid propellant rocket would be fired to provide the final velocity increment necessary for orbital insertion and circularization. Payloads launched (or perhaps more correctly "shot") into orbit by such electromagnetic guns would experience peak accelerations ranging from hundreds to thousands of g's (where one g is the normal acceleration due to gravity at Earth's surface).

GLTS has also been suggested for use on the lunar surface, where the absence of an atmosphere and the Moon's reduced gravitational field (about one-sixth that experienced on Earth's surface) might make this launch approach attractive for hurling raw or refined materials into low lunar orbit.

Launch concepts involving this approach are sometimes referred to as the "Jules Verne approach to orbit"—in recognition of the giant gun used by the French writer Jules Verne (1828–1905) to send his explorers on a voyage around the Moon in the famous science fiction story *From the Earth to the Moon.*

✧ Laser Propulsion System

The laser propulsion system concept is a suggested form of advanced in space propulsion in which a high-energy laser beam is used to heat a propellant (working fluid) to extremely high temperatures and then the heated propellant is expanded through a nozzle to produce thrust.

If the high-energy laser is located away from the space vehicle—as, for example, on Earth, on the Moon, or perhaps at a different orbital position in space—the laser energy used to heat the propellant would be beamed into the space vehicle's thrust chamber. The beamed energy propulsion concept offers the advantage that the space vehicle being propelled need not carry a heavy onboard power supply. A fleet of laser-propelled orbital transfer vehicles operating between low Earth orbit (LEO) and lunar orbit has been suggested as part of a lunar-based transportation system. Aerospace engineers have also suggested the application of the beamed energy propulsion concept for deep space probes that travel on one-way missions to the edges of the solar system and beyond.

In an interesting variant of the beamed energy propulsion concept, the incoming laser energy (that is, "light energy") might first be gathered by a large optical collector onboard the space probe and then converted into electric energy, which is then used to operate some type of advanced electric propulsion system. The onboard electric propulsion system would accommodate more precise maneuvering as the deep space probe encountered and explored various trans-Neptunian objects of scientific interest.

✧ Solar Sail

The solar sail is a proposed method of in-space transportation that uses solar radiation pressure to gently push a giant gossamer structure and its payload through interplanetary space.

As presently envisioned, the solar sail would use a large quantity of very thin, low-mass (ideally about 6.42×10^{-5} lb/ft^2 [1.0×10^{-3} kg/m^2]) reflective material to produce a net reaction force by reflecting incident sunlight. Because solar radiation pressure is very weak and decreases as the square of the distance from the Sun, enormous sails—perhaps 1.08 million square feet (100,000 m^2) to 5.38 million square feet (500,000 m^2)—would be needed to achieve useful accelerations and payload transport.

The main advantage of the solar sail would be its long-duration operation as an interplanetary transportation system. Unlike rocket propulsion systems that must expel their onboard supply of propellants to generate thrust, solar sails have operating times only limited by the effective lifetimes in space of the sail materials. The solar photons that do the "pushing" constantly pour in from the Sun and are essentially "free." This makes the concept of solar sailing particularly interesting for cases where space mission planners must ship massive, but nonpriority, payloads through interplanetary space—as, for example, a shipment from Earth to the Mars base of spare parts for robot exploration vehicles.

Unfortunately, because the large reflective solar sail cannot generate a force opposite to the direction of the incident solar radiation flux, its maneuverability is limited. This lack of maneuverability, along with long transit times, represents the major disadvantages of the solar sail as an in-space transportation system.

✧ Plasma Sail

Another promising sail concept now being investigated by NASA for in space propulsion is the plasma sail. The plasma sail is a huge, magnetic bubble generated aboard a small interplanetary vehicle and pushed along by interactions with charged particles in the solar wind. Space scientists estimate that plasma sail technologies, with negligible mass and benefit-ing from an inexhaustible supply of solar wind particles from the Sun, could cut conventional trip times to the outer planets in half. Solar wind particles travel outward from the Sun at speeds up to 28 miles per second (45 km/s).

✧ Aerocapture

Aerocapture is the use of a planet's atmosphere to slow down a space-craft. It is part of a unique family of "aeroassist" technologies that will support missions to the most distant planets of our solar system. A space vehicle with aerocapture technology integrated in its design approaches a target planet on a hyperbolic trajectory. Aerocapture technology enables the spacecraft to be quite literally "captured" into orbit as it passes at the proper encounter angle through the planet's atmosphere. The spacecraft capture takes place without the use of retrorockets and the subsequent consumption of onboard propellant supplies. This fuel-free capture tech-nique can reduce the mass of a typical interplanetary spacecraft by half. Aerocapture allows aerospace engineers to build a smaller and less-expen-sive scientific spacecraft. Spacecraft designers and mission planners often use the saved "propellant mass" allocation to build a better-equipped and more robust spacecraft—greatly improving the long-term science pro-gram at and around the target planet.

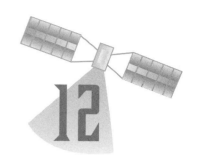

Interstellar Probes and Starships

This chapter takes an over-the-technology horizon peek at some conceptual propulsion systems that now lie far beyond today's level of aerospace engineering, or even beyond very generous extrapolations of foreseeable aerospace developments through the first half of this century. Why bother with such futurism? The reason is quite simple. A little over a century ago, the great technical dream that inspired space travel advocates was the vision of interplanetary travel. That long-cherished dream has now come true and in its place there is another inspiring vision—the dream of interstellar travel.

Science fiction writers and rocket scientists alike recognize that interstellar travel is basically matter transport between star systems in a galaxy—specifically, the ability to send material objects and possible people beyond our own solar system. Upon closer inspection, interstellar travel falls into several interesting categories. The first category involves the one-way journey of robot interstellar probe on a pioneering mission of exploration. The plaques carried on NASA's *Pioneer 10* and *11* spacecraft and the "recorded message" found on NASA's *Voyager 1* and *2* spacecraft are very simple examples. While these particular 20th-century robot spacecraft are now on interstellar trajectories, they are traveling at such incredibly slow speeds that it will take them millions of years of drifting through the interstellar void before they get in the general neighborhood of some of the closest star systems.

The next category of interstellar travel involves a somewhat more sophisticated robot spacecraft that possesses a high level of artificial (machine) intelligence and also carries an extensive summary of the cultural and technical heritage of its parent civilization. This very smart robot probe would be capable of repairing itself, or possibly even replicating itself, using materials found within alien solar systems. Once sent out into the interstellar void, this advanced robot interstellar probe creates a ripple or wave of exploration that, depending on physical circumstances, might propagate an appreciable distance into a galaxy away from the parent civilization. Time is a key factor here. As the intelligent robot–exploring machine and any mechanical progeny it may construct wander for millennia through interstellar space, they would visit and explore many new worlds around distant suns. Over this extended period of time, such robot probes would continue to send reports back to its parent civilization—with each report taking a bit longer to arrive as the probe gets many light years away. Ultimately, the parent (or launching) civilization becomes extinct and these self-repairing interstellar probes become the legacy of their existence within the galaxy.

The third category of interstellar travel involves a starship crewed by intelligent (biological) beings on a long-term round-trip voyage of scientific exploration to one or several nearby star systems. One premise inherent in such extended crewed voyages of exploration is that the descendants (children, grandchildren, great grandchildren, etc.) of the original crew return to the parent solar system after a few centuries. As discussed a little later in this chapter, the relativistic effects of time dilation could produce some interesting social circumstances.

The final category of interstellar travel would involve a very large starship, or giant interstellar ark, that is designed to transport a self-sufficient portion of the human race on a one-way mission beyond the solar system in search of new, suitable planetary systems to explore and inhabit. How far humankind's future descendants would get into the Milky Way Galaxy is a matter of great speculation. But once departed from the solar system, this chunk of humanity would begin to spread intelligent (biological) life around the universe.

The underlying need to make any or all of these interstellar travel categories possible is energy, enormous quantities of energy for propulsion. The remainder of this chapter examines some very speculative (when viewed within the context of contemporary rocket science) candidate propulsion options to take spacecraft beyond the boundaries of our solar system.

✧ Interstellar Probe

An interstellar probe is a highly automated robotic spacecraft sent from one solar system to explore other star systems. In all likelihood, this type of probe would involve a level of machine intelligence capable of operating the spacecraft autonomously for decades or even centuries. Once the robot probe arrives at a new star system, it would initiate a detailed exploration protocol. For example, the target star system is scanned for possible life-bearing planets, and if any are detected, they become the object of more intense scientific investigations. Data may also be gathered with a series of mini-probes—deployed from the main robot spacecraft to investigate individual objects of interest within the new star system. Interesting scientific discoveries would be transmitted back to the home civilization (in our case, Earth) via telecommunication equipment carried by the main robot probe, serving as a mother spacecraft. After light-years of travel, these signals would be intercepted and analyzed by scientists on Earth and any discoveries and used to enrich humankind's understanding of the universe.

Robot interstellar probes also might be designed to carry specially engineered microorganisms, spores, and bacteria. If a probe encounters ecologically suitable planets on which life has not yet evolved, then it could "seed" such barren but potentially fertile worlds with primitive life-forms

This artist's rendering shows how the first interstellar robot probe might look as it departs the solar system (circa 2075) on an epic journey of scientific exploration. *(NASA)*

RELATIVITY

Relativity is the theory of space and time developed by Albert Einstein (1879–1955), which has become one of the foundations of modern physics. Einstein's theory of relativity often is discussed in two general categories: the special theory of relativity, which he first proposed in 1905, and the general theory of relativity, which he presented starting in 1915.

The special theory of relativity is concerned with the laws of physics as seen by observers moving relative to one another at constant velocity—that is, by observers in nonaccelerating, or inertial, reference frames. Special relativity has been well demonstrated and verified by many types of experiments and observations.

Einstein proposed two fundamental postulates in formulating special relativity. The first postulate of special relativity is: The speed of light (c) has the same value for all (inertial-reference-frame) observers, regardless and independent of the motion of the light source or the observers. The second postulate of special relativity is: All physical laws are the same for all observers moving at constant velocity with respect to each other.

The first postulate appears contrary to our everyday "Newtonian mechanics" experience. Yet the principles of special relativity have been more than adequately validated in experiments. Using special relativity, scientists can now predict the space-time behavior of objects traveling at speeds from essentially zero up to those approaching that of light itself. At lower velocities the predictions of special relativity become identical with classical Newtonian mechanics. However, when we deal with objects moving close to the speed of light, we must use relativistic mechanics.

What are some of the consequences of the theory of special relativity? The first interesting relativistic effect is called time dilation. Simply stated—with respect to a stationary observer/clock—time moves more slowly on a moving clock/system. This unusual relationship is described by the equation

$$\Delta t = (1/\beta)\ \Delta T_p$$

where Δt is called the time dilation (the apparent slowing down of time on a moving clock relative to a stationary clock/observer) and ΔT_p is the "proper time" interval as measured by an observer/clock on the moving system, and β is defined as

$$\beta = \sqrt{[1 - (v^2/c^2)]}$$

where v is the velocity of the object and c is the velocity of light.

Consider the time-dilation effect with respect to a postulated starship flight from the solar system. The scenario starts with twin brothers, Astro and Cosmo, who are both astronauts and are currently 25 years of age. Astro is selected for a special 40-year-duration starship mission, while Cosmo is selected for the ground control team. This particular starship, the latest in the fleet, is capable of cruising at 99 percent of the speed of light (0.99 c) and can quickly reach this cruising speed. During the mission, Cosmo, the twin who stayed behind on Earth, ages 40 years. (The assumption here is that Earth is the fixed or stationary reference frame "relative" to the starship.) Due to time-dilation, Astro—the twin who has been on board the starship cruising the galaxy at 99 percent of the speed of light for the last 40 Earth-years—has aged just 5.64 Earth-years! When he returns to Earth from the starship mission, he is a little over 30 years old, while his twin brother, Cosmo, is now 65 and retired in Florida. Obviously, starship travel (if scientists can overcome some extremely challenging technical barriers) also presents some very interesting social problems.

The time-dilation effects associated with near-light speed travel are real. Scientists have observed and measured such phenomena in a variety of modern experiments. All physical processes (chemical reactions, biological processes, nuclear-decay phenomena and so on) appear to slow down when in motion relative to a "fixed" or stationary observer/clock.

Another interesting effect of relativistic travel is length contraction. Scientists define an object's proper length (L_p) as its length measured in a reference frame in which the object is at rest. Then, the length of the object when it is moving (L)—as measured by a stationary observer—is always smaller, or contracted. Physicists use the following equation to describe relativistic length contraction:

$$L = \beta (L_p)$$

This apparent shortening, or contraction, of a rapidly moving object is seen by an external observer (in a different inertial reference frame) only in the object's direction of motion. In the case of a starship traveling at near-light speeds, to observers on Earth this vessel would appear to shorten, or contract, in the direction of flight. If an alien starship was 0.62 mile (1 km) long (at rest) and entered the solar system at an encounter velocity of 90 percent of the speed of light (0.9 c), then a terrestrial observer would see a starship that appeared to be about 0.27 mile (0.435 km) long. The aliens on board and all their instruments (including tape measures) would look contracted to external observers, but would not appear any shorter to those on board the ship—that is, to observers within the moving reference frame.

If this hypothetical alien starship were really "burning rubber" at a velocity of 99 percent of the speed of light (0.99 c), then its apparent contracted length to an observer on Earth would be about 0.088 mile (0.141 km) or 462 feet (141 m).

In contrast, if this alien starship was just a "slow" interstellar freighter that was lumbering along at only 10 percent of the speed of light (0.1 c), then it would appear to be about 0.618 mile (0.995 km) long to an observer on Earth.

Special relativity also influences the field of dynamics. Although the rest mass (m_o) of a body is invariant (does not change), its "relative" mass increases as the speed of the object increases with respect to an observer in another fixed or inertial reference frame. An object's relative mass is given by:

$$m = (1/\beta) \, m_o$$

This simple equation has far-reaching consequences. As an object approaches the speed of light, its mass becomes infinite. Since things cannot have infinite masses, physicists conclude that material objects cannot reach the speed of light. This is basically the "speed-of-light barrier," which appears to limit the speed at which interstellar travel can occur.

From the theory of special relativity, scientists now conclude that only a "zero-rest-mass" particle, such as a photon, can travel at the speed of light. There is one other major consequence of special relativity that has greatly affected our daily lives—the equivalence of mass and energy from Einstein's very famous formula:

$$E = \Delta m \, c^2$$

where E is the energy equivalent of an amount of matter (Δm) that is annihilated or converted completely into pure energy and c is the speed of light. This simple yet powerful equation explains where all the energy in nuclear fission or nuclear fusion comes from.

In 1915, Einstein introduced his general theory of relativity. He used this development to describe the space-time relationships developed in special relativity for cases where there was a

(continues)

(continued)

strong gravitational influence such as white dwarf stars, neutron stars, and black holes. One of Einstein's conclusions was that gravitation is not really a force between two masses (as postulated in Newtonian mechanics) but rather arises as a consequence of the curvature of space-time. In a four-dimensional universe (x, y, z, and time), space-time becomes curved in the presence of matter—especially very massive, compact objects.

The fundamental postulate of general relativity is also called Einstein's principle of equivalence: The physical behavior inside a system in free-fall is indistinguishable from the physical behavior inside a system far removed from any gravitating matter (that is, the complete absence of a gravitational field).

Several experiments have been performed to confirm the general theory of relativity. These experiments have included observation of the bending of electromagnetic radiation (starlight and radio wave transmissions from various spacecraft missions, such as NASA's *Viking* spacecraft on Mars) by the Sun's immense gravitational field and the recognition of the subtle perturbations (disturbances) in the orbit of the planet Mercury (at perihelion—the point of closest approach to the Sun) as caused by the curvature of space-time in the vicinity of the Sun. While some scientists do not think that these experiments have conclusively demonstrated the validity of general relativity, additional astronomical measurements—made with more powerful space-based observatories that investigate phenomena, such as neutron stars and black holes—continue to collect evidence supporting general relativity. Special scientific spacecraft, such as NASA's *Gravity Probe B*, are also examining phenomena associated with general relativity.

or at least life precursors. In that way, human beings (through their robot probes) would not only be exploring neighboring star systems, but would also be participating in the spread of life in the Milky Way Galaxy.

A self-replicating system (SRS) is a truly advanced interstellar robot probe. The Hungarian-American mathematician John von Neumann (1903–57) was the first person to seriously consider the problem of self-replicating machine systems. During and following World War II, he became interested in the study of machine replication as part of his wide-ranging interests in complicated machines. A single SRS unit is a machine system that contains all the elements required to maintain itself, to manufacture desired products, and even (as the name implies) to reproduce itself. From von Neumann's initial work and the more recent work of other investigators, five general classes of SRS behavior have been defined: production, replication, growth, repair, and evolution.

The issue of closure (total self-sufficiency) is one of the fundamental problems in designing self-replicating systems. In an arbitrary SRS unit there are three basic requirements necessary to achieve total closure: matter closure, energy closure, and information closure. If the machine device is only partially self-replicating, then it is said that only partial closure of the system has occurred. In this case, some essential matter, energy or information must be provided from external sources, or else the machine would fail to reproduce itself.

Within this solar system, a self-replicating system could be used to assist in planetary engineering projects—where a seed machine would be sent to a target planet, like Mars. The SRS unit would make a sufficient number of copies of itself (using native Martian resources, for example) and then set about some production task, like manufacturing oxygen to make the planet's atmosphere more breathable for future human settlers. Similarly, a seed SRS unit could be sent into interstellar space and trigger a wave of galactic exploration—stopping to repair itself or possibly even make copies of itself in the various alien solar systems it encounters on its cosmic journey. In theory a single successful SRS interstellar probe—capable of traveling at perhaps 70 percent of the speed of light—could propagate a wave of smart exploring machines that spreads through the entire galaxy in a few hundred million years.

✧ Starships

A starship is a space vehicle capable of traveling the great distances between star systems. Even the closest stars in the Milky Way Galaxy are typically several light-years apart. By convention, the word *starship* is used here to describe interstellar spaceships capable of carrying intelligent beings to other star systems; robot interstellar spaceships are called interstellar probes.

What are the performance requirements for a starship? First, and perhaps most important, the vessel should be capable of traveling at a significant fraction of the speed of light (c). Ten percent of the speed of the light ($0.1 c$) is often considered as the lowest acceptable speed for a starship, while cruising speeds of $0.9 c$ and beyond are considered highly desirable. This optic-velocity cruising capability is necessary to keep interstellar voyages to reasonable lengths of time, both for the home civilization and for the starship crew.

Consider, for example, a trip to the nearest star system, Alpha Centauri—a triple star system about 4.23 light-years away. At a cruising speed of $0.1 c$, it would take about 43 years just to get there and another 43 years to return. The time-dilation effects of travel at these "relatively low" relativistic speeds would not help too much either, since a ship's clock would register the passage of about 42.8 years versus a terrestrial ground elapse time of 43 years. In other words, the crew would age about 43 years during the journey to Alpha Centauri. If we started with 20-year-old crewmembers departing from the outer regions of the solar system in the year 2100 at a constant cruising speed of 0.1 c, they would be approximately 63 years old when they reached the Alpha Centauri star system some 43 years later in 2143. The return journey would be even more dismal. Any surviving crewmembers would be 106 years old when the ship returned to the solar system in the year 2186. Most if not all the crew would probably have died of old age or boredom. And that's for just a journey to the nearest star.

A starship should also provide a comfortable living environment for the crew and passengers, especially in the case of an interstellar ark. Living in a relatively small, isolated and confined habitat for a few decades to perhaps a few centuries can certainly overstress even the most psychologically adaptable individuals and their progeny. One common technique used in science fiction to avoid this crew stress problem is to have all or most of the crew placed in some form of suspended animation, while the vehicle travels through the interstellar void, tended by a ship's company of smart robots.

Any properly designed starship must also provide an adequate amount of radiation protection for the crew, passengers, and sensitive electronic equipment. Interstellar space is permeated with galactic cosmic rays. Nuclear radiation leakage from an advanced thermonuclear fusion engine or a matter-antimatter engine (photon rocket) must also be prevented from entering the crew compartment. In addition, the crew will have to be protected from nuclear radiation showers produced when a starship's hull, traveling at near light speed, slams into interstellar molecules, dust or gas. For example, a single proton (which we can assume is "stationary" relative to the starship) being hit by a starship moving at 90 percent of the speed of light (0.9 c) would appear to those on board like an enormously energetic, gigaelectron volt (GeV)–class proton being accelerated at them. Imagine traveling for years at the beam output end of a very-high-energy particle accelerator. Without proper deflectors or shielding, survival in the crew compartment from such nuclear radiation doses is very doubtful.

To truly function as a starship, the vessel must be able to cruise at will, light-years from its home star system. The starship must also be able to accelerate to significant fractions of the speed of light; cruise at these near-optic velocities; and then decelerate to explore a new star system or to investigate a derelict alien spaceship found adrift in the depths of interstellar space.

It is beyond the scope of this book to discuss the difficulties of navigating through interstellar space at near light velocities. It is sufficient just to mention here that when crew "looks" forward at near light speeds everything is blueshifted, while when they look aft (backward) things appear redshifted. This occurs because of the Doppler effect. In order to venture where no one else has gone before, the starship crew must be able to find their way from one location in the Milky Way to another galaxy totally on their own.

What appears to be the major engineering technology needed to make the starship a real part of human history is an effective propulsion system. Interstellar-class propulsion technology is the key to the galaxy for any intelligent species that has mastered spaceflight within and up to the limits of its own solar system. Despite the tremendous engineering difficulties associated with the development of a starship propulsion system, several basic and quite easily grasped physical concepts have been proposed. These include the pulsed nuclear fission engine (Project Orion concept), the pulsed nuclear fusion concept (Project Daedalus study), the interstellar nuclear ramjet, and the photon rocket. The principles of operation associated with each of these conceptual systems are now discussed along with some of their potential advantages and disadvantages.

✦ Pulsed Nuclear Fission Engine (Project Orion Concept)

Project Orion was the name given to a nuclear-fission pulsed rocket concept studied by agencies of the U.S. government in the early 1960s. Exploding a series of nuclear-fission devices behind a human-crewed interplanetary spaceship would propel it. A giant pusher plate, mounted on large shock absorbers, receives the energy pulse from each successive nuclear detonation, and the spaceship configuration would be propelled forward by Newton's action-reaction principle. In theory, the pulsed nuclear fission engine might achieve specific impulse values ranging from 2,000 to 6,000 seconds (19,600 to 58,800 m/s in the SI unit system)—depending on the size of the pusher plate and the number and yield of the nuclear explosions.

A crewed Project Orion spaceship would move rapidly throughout interplanetary space at a steady acceleration of perhaps 0.5 g (one-half the acceleration of gravity on Earth's surface). Typically, a 1- to 10-kiloton fission device would be exploded every second or so close behind the giant pusher plate. A kiloton is the energy of a nuclear explosion that is equivalent to the detonation of 1,000 tons of TNT (that is, the chemical high-explosive trinitrotoluene). (In the American engineering unit system, a ton is defined as 2,000 pounds-mass [909 kg].) Studies suggest that a

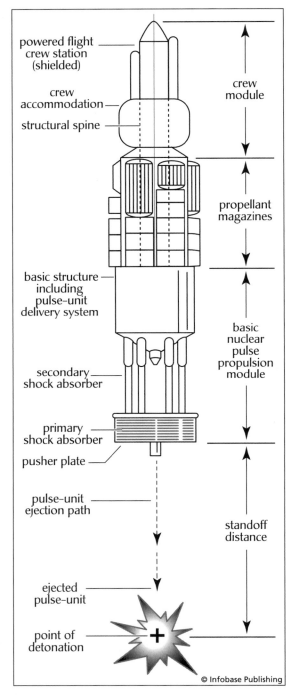

powered flight crew station (shielded)

crew accommodation

structural spine

crew module

basic structure including pulse-unit delivery system

propellant magazines

basic nuclear pulse propulsion module

secondary shock absorber

primary shock absorber

pusher plate

pulse-unit ejection path

standoff distance

ejected pulse-unit

point of detonation

© Infobase Publishing

Major elements of the original Project Orion pulsed nuclear fission spaceship concept

few thousand such detonations would be needed to propel a crew of 20 astronauts to Mars or the moons of Jupiter.

Work by the United States on the nuclear-fission pulse rocket concept came to an end in the mid-1960s, as a result of the Limited Test Ban Treaty of 1963. This treaty prohibited the signatory nations (which included the United States) from testing nuclear devices in Earth's atmosphere, underwater, or in outer space. Since then, other versions of the original Project Orion concept have emerged. In these new concepts, many small, controlled thermonuclear-fusion explosions—taking place inside a specially constructed thrust chamber—replace externally detonated nuclear fission explosions. These mini-thermonuclear explosions might occur in an inertial confinement fusion (ICF) process in which many powerful laser, electron, or ion beams simultaneously impinge on a tiny fusion pellet. Each miniature thermonuclear explosion would have an explosive yield equivalent to a few tons of TNT. The expanding shell of very hot, ionized gas from the thermonuclear explosion would be directed into a thrust-producing exhaust stream. (For additional discussion of this concept, see Project Daedalus in the next section.)

Project Orion-type, pulsed nuclear fission spaceships, if ever developed, could open up the entire solar system to exploration by human crews. For example, Earth-to-Neptune travel could take less than 15 days at a steady, comfortable constant acceleration of one g. Pulsed-nuclear fission systems also represent a candidate propulsion system for interstellar travel—but not a very robust one.

One drawback of the pulsed nuclear fission system is the relatively low efficiency by which it converts propellant (explosive device) mass into useful energy for propulsion. Furthermore, there are only a limited number of nuclear explosive devices that could be carried onboard and the heavy shielding is needed to reduce the nuclear radiation hazards to the crew to acceptable levels. Although easy to visualize and even tested (in nonnuclear prototype models), interstellar travel using the Project Orion concept is probably limited to a maximum speed of about 1 percent to 10 percent the speed of light. Therefore, this type of advanced propulsion system remains more useful for rapid interplanetary missions to the far reaches of our own solar system. However, the concept may have a very limited role in early interstellar probe missions. Furthermore, the pulse nuclear fission system could also find limited application as an auxiliary (or emergency) propulsion system for a huge, but relatively slow, interstellar ark that is making a centuries-long interstellar voyage.

✧ Pulsed Nuclear Fusion System (Project Daedalus)

Project Daedalus is the name given to an extensive study of interstellar space exploration conducted from 1973 to 1978 by a team of scientists and engineers under the auspices of the British Interplanetary Society. This hallmark effort examined the feasibility of performing a simple interstellar probe mission using only contemporary technologies and/or reasonable extrapolations of imaginable near-term capabilities.

In mythology, Daedalus was the grand architect of King Minos's labyrinth for the Minotaur on the island of Crete. But Daedalus also showed the Greek hero Theseus, who slew the Minotaur, how to escape from the labyrinth. An enraged King Minos imprisoned both Daedalus and his son, Icarus. Undaunted, Daedalus (a brilliant engineer) fashioned two pairs of wings out of wax, wood, and leather. Before their aerial escape from a prison tower, Daedalus cautioned his son not to fly too high, so that the Sun would not melt the wax and cause the wings to disassemble. They made good their escape from King Minos's Crete, but while over the sea Icarus—an impetuous teenager—ignored his father's warnings and soared high into the air. Daedalus (who reached Sicily safely) watched in horror as Icarus's wings collapsed and the young man tumbled to his death in the sea below.

The proposed Daedalus interstellar probe structure, communications systems, and much of the payload were designed entirely within the parameters of 20th-century technology. Other components, such as the advanced machine intelligence flight controller and onboard computers for in-flight repair, required artificial-intelligence capabilities expected to be available in the mid-21st century. The propulsion system, perhaps the most challenging aspect of any interstellar mission, was designed as a nuclear-powered, pulsed-fusion rocket engine that burned an exotic thermonuclear fuel mixture of deuterium and helium-3 (a rare isotope of helium). This pulsed-fusion system was believed capable of propelling the robot interstellar probe to velocities in excess of 12 percent of the speed of light. The best source of helium-3 was considered to be the planet Jupiter. So, one of the major technologies recommended for Project Daedalus was an ability to mine the Jovian atmosphere for helium-3. This mining operation might be achieved by using "aerostat" extraction facilities (floating balloon-type factories).

The Project Daedalus engineering team also suggested that this ambitious interstellar flyby (one-way) mission could be undertaken at the end of the 21st century—when the successful development of humankind's extraterrestrial civilization had generated the necessary wealth, technology base, and exploratory zeal. The target selected for this first interstellar probe was Barnard's star—a red dwarf (spectral type M star) about 5.9 light-years away in the constellation Ophiuchus.

The Daedalus spaceship would be assembled in cislunar space (partially fueled with deuterium from Earth) and then ferried to an orbit around Jupiter, where it could be fully fueled with the helium-3 propellant that had been mined out of the Jovian atmosphere. These thermonuclear fuels would then be prepared as pellets, or fusion "targets," for use in the ship's two-stage pulsed-fusion power plant. Once fueled and readied for its epic interstellar voyage, somewhere around the orbit of Callisto, the ship's mighty pulsed-fusion first-stage engine would come alive. This first-stage pulsed-fusion unit would continue to operate for about two years. At first-stage shutdown, the vessel would be traveling at about 7 percent of the speed of light.

The expended first-stage engine and fuel tanks would be jettisoned in interstellar space, and the second-stage pulsed-fusion engine would ignite. The second stage would also operate in the pulsed-fusion mode for about two years. Then, it too would fall silent. At that point, the giant robot spacecraft, with its cargo of sophisticated remote sensing equipment and nuclear fission-powered probe ships, would be traveling at approximately 12 percent of the speed of light. The Daedalus spaceship would then need about 47 years of coasting (after second-stage shutdown) to encounter Barnard's star.

In this scenario, when the interstellar probe is about three light-years away from its objective (after 25 years mission elapsed time), smart computers on board would initiate long-range optical and radio astronomy observations. A special automated effort would be made to locate and identify any extrasolar planets or other interesting celestial objects that exist in the Barnardian system.

Traveling at 12 percent of the speed of light, Daedalus would only have a very brief physical passage through the target star system. The long-awaited encounter actually amounts to a few days of "close-range" observation of Barnard's star and only "minutes" of observation of any planets or other interesting objects by instruments onboard the large robot mother spacecraft.

Several years before the Daedalus mother spacecraft zips through the Barnardian system, it would launch its complement of nuclear-powered probes (also traveling at 12 percent of the speed of light initially). These probe ships, individually targeted to objects of potential interest by computers onboard the robot mother spacecraft, would fly ahead and serve as data-gathering scouts. The Project Daedalus team considered a complement of 18 small robot probe spacecraft appropriate for the mission.

As the Daedalus mother spaceship flashed through the Barnardian system, it would gather data from its own onboard instruments, as well as information relayed to it by the scout spacecraft. Over the next day or so, the mother spacecraft would transmit all these mission data back to Earth, where team scientists would patiently wait the approximately six years it takes for the information-laden electromagnetic waves to cross the interstellar void, traveling at light speed.

Its mission completed, the Daedalus mother spaceship minus its probes would continue on a one-way journey into the darkness of the interstellar void—to be discovered perhaps millennia later by an advanced alien race, which might puzzle over humankind's first attempt at the direct exploration of another star system.

The main conclusions drawn from the Project Daedalus study are summarized as follows: (1) exploration missions to other star systems are, in principle, technically feasible; (2) missions of this type would provide scientists a great deal of new information about the origin, extent, and physics of the Milky Way galaxy, as well as the formation and evolution of stellar and planetary systems; (3) the prerequisite interplanetary and initial interstellar space system technologies necessary to successfully conduct this class of mission also contribute significantly to the search for extraterrestrial intelligence (for example, smart robot probes and interstellar communications); (4) a long-range societal commitment on the order of a century would be required to achieve such a project; and (5) the prospects for interstellar flight by human beings do not appear very promising, using current or foreseeable 21st-century space technologies.

The Project Daedalus study also identified three key technology advances that are necessary to make even a one-way robot interstellar mission possible. These are (1) the development of controlled nuclear fusion, especially the use of the deuterium/helium-3 thermonuclear reaction; (2) advanced machine intelligence; and (3) the ability to extract helium-3 in large quantities from the Jovian atmosphere, or other sources in the outer solar system.

Although the choice of Barnard's star as the target for the first interstellar mission was somewhat arbitrary, if future human generations can build such an interstellar robot spaceship and successfully explore the Barnardian system, then with modest technology improvements, all star systems within 10 to 12 light-years of Earth become potential targets for a more ambitious program of interstellar exploration using robot interstellar probes.

✧ Interstellar Ramjet

This giant starship is first accelerated to near–light speed by some other propulsion technique. Then, a giant scoop (perhaps a hundred or so

An artist's rendering of an interstellar ramjet. The conceptual starship is characterized by a huge (100–square-kilometer) scoop to capture interstellar hydrogen for use as fuel in sustaining the thermonuclear reactions that propel the vehicle to upward of 90 percent of light speed. *(NASA/Marshall Space Flight Center)*

square kilometers in area) collects interstellar hydrogen, which fuels a proton-proton thermonuclear cycle. The energy-liberating thermonuclear reaction products exit the vehicle and produce forward thrust (see figure on page 180). The interstellar ramjet concept is sometimes referred to as the Bussard ramjet, after the American engineer Robert Bussard, who suggested this concept in 1960.

In principle, the performance of the interstellar ramjet would not be limited by the amount of propellant that it could carry. However, the construction of a durable, low-mass, giant scoop presents a major technical hurdle. Conceptually capable of producing speeds from 10 percent to 90 percent that of light, the interstellar ramjet would certainly be useful to propel interstellar probes, starships, and giant space arks. However, this interstellar propulsion concept requires many major technical breakthroughs, including that of controlled nuclear fusion using the proton-proton cycle that place in some stellar interiors. Therefore, from the sobering perspective of early 21st-century engineering and physics, the development of an interstellar ramjet appears several centuries away, if ever.

✧ Photon Rocket

This type of interstellar propulsion system uses matter and antimatter as propellant. Equal amounts are combined and annihilate each other, releasing an equivalent amount of energy in the form of hard nuclear (gamma) radiation. In the ultimate version of this conceptual propulsion system, these gamma ray photons are focused and emitted in a collimated beam out the back of the vessel, providing a forward thrust. In a less efficient intermediate photon rocket concept, the energy released in matter-antimatter reactions is used to heat and accelerate a working fluid, which then propels the vehicle.

The photon rocket represents the best theoretical propulsion system contemporary physics permits. This exotic reaction engine offers cruising speeds ranging from 10 percent to 99 percent of the speed of light. Such conceptual performance finds application in interstellar probes (including self-replicating systems), starships, and large interstellar arks. While physicists have produced and studied limited quantities of antimatter (primarily in the form of antihydrogen), enormous engineering barriers must be overcome before operationally useful quantities of antihydrogen are produced and stored as propellant for an interstellar space vehicle. The efficient collimation of annihilation radiation (gamma-ray photons) and radiation shielding represent other enormous technical hurdles.

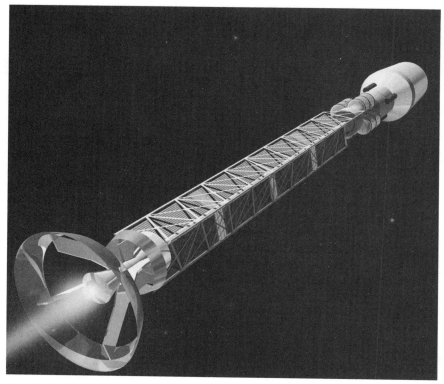

An artist's rendition of an antimatter propulsion system, or photon rocket. In principle, matter–antimatter annihilation offers the highest possible physical energy density of any known reaction substance. *(NASA/Marshall Space Flight Center)*

✧ Hyperspace, Wormholes, and Manipulation of Space-Time Continuum

The final option for interstellar travel discussed here currently resides almost exclusively in the realm of science fiction—although contemporary research in physics is providing some tantalizing hints that 22nd-century science may be significantly different from today's understanding of the physical universe. Spaceships capable of manipulating the space-time continuum (artistically rendered in the figure on page 183) could prove to be the key to interstellar travel.

Hyperspace is a concept of convenience developed in science fiction to make "faster-than-light" travel appear credible. This hypothetical concept is frequently described as a special dimension or property of the universe in which physical things are much closer together than they are in the normal space-time continuum. In a typical science fiction story, the crew-members of a spaceship simply switch into hyperspace drive, and distances to objects in the "normal" universe are considerably shortened. When the spaceship emerges out of hyperspace, the crew is where they wanted to be, essentially instantly. Although this concept violates the speed-of-light

A highly speculative artist's rendering of a starship that manipulates the space-time continuum to achieve travel at luminal, or perhaps superluminal, speeds. This relativistic "fire arrow" would take advantage of principles and phenomena that are now only vaguely being hinted at in contemporary physics. The way such a starship could successfully interact with space-time to perform interstellar travel is graciously left as a challenge to the rocket engineers of tomorrow—or perhaps many tomorrows beyond tomorrow. *(NASA/Marshall Space Flight Center)*

barrier predicted by Albert Einstein's special relativity theory, it is nevertheless quite popular in modern science fiction.

In another interesting speculation, some scientists conjecture that matter falling into a black hole may actually survive. They further suggest that under very special circumstances such matter might be conducted by means of passageways, called wormholes, to emerge in another place or time in this universe or perhaps even in another universe. These hypothetical wormholes are considered to be distortions in the space-time continuum. In entertaining this line of thinking, scientists are suggesting that black holes play "relativistic tricks" with space-time. If wormholes really do exist, then, in principle (at least), a person might be able to use one to travel faster than light. Use of a wormhole might allow a starship's crew to visit distant parts of the universe—or possibly to travel through time as well as through space.

Conclusion

I n the middle of the 20th century, developments in rocket technology revolutionized warfare, transformed international politics, and changed forever the trajectory of human civilization. One of the earliest milestones involved the combination of two powerful World War II–era weapons systems, the American nuclear bomb and the German V-2 ballistic missile. Cold war politics encouraged a hasty marriage of these emerging military technologies. The union ultimately produced an offspring called the intercontinental ballistic missile (ICBM)—a weapon system that most military experts regard as the single most influential weapon in all of history.

Modern rockets are also sophisticated propulsion machines that harness the action-reaction principle to provide access to outer space. Rockets used as space launch vehicles have enabled humankind to leave planet Earth and explore the solar system. Human-crafted space vehicles have visited all the major planets (soon including tiny Pluto) and reconnoitered numerous other interesting celestial bodies. Continued advances in rocket technology this century will greatly expand the scientific investigation of the solar system, assist in the search for life, return human beings to the Moon as permanent settlers, and then carry the first human explorers to Mars and beyond. Projections of contemporary rocket technologies offer the entire solar system to the human race as a destination and a destiny.

Travel through interstellar space appears to present some very formidable barriers and unprecedented technical challenges. By the end of this century, however, human beings may elect to tackle some of these challenges and make the social and technical commitments necessary to explore nearby star systems with robot probes. This decision would represent a major milestone in the emergence of human beings as an intelligent species within the galaxy. The ancient Roman phrase, *Ad astra per aspera,* sums up the opportunity and challenges ahead; it means "to the stars through persevering hard work."

Chronology

✧ **ca. 3000 B.C.E. (to perhaps 1000 B.C.E.)**
Stonehenge erected on the Salisbury Plain of Southern England (possible use: ancient astronomical calendar for prediction of summer solstice)

✧ **ca. 1300 B.C.E.**
Egyptian astronomers recognize all the planets visible to the naked eye (Mercury, Venus, Mars, Jupiter, and Saturn), and they also identify over 40 star patterns or constellations

✧ **ca. 500 B.C.E.**
Babylonians devise zodiac, which is later adopted and embellished by Greeks and used by other early peoples

✧ **ca. 375 B.C.E.**
The early Greek mathematician and astronomer Eudoxus of Cnidos starts codifying the ancient constellations from tales of Greek mythology

✧ **ca. 275 B.C.E.**
The Greek astronomer Aristarchus of Samos suggests an astronomical model of the universe (solar system) that anticipates the modern heliocentric theory proposed by Nicolaus Copernicus. However, these ideas, which Aristarchus presents in his work *On the Size and Distances of the Sun and the Moon,* are essentially ignored in favor of the geocentric model of the universe proposed by Eudoxus of Cnidus and endorsed by Aristotle

✧ **ca. 129 B.C.E.**
The Greek astronomer Hipparchus of Nicaea completes a catalog of 850 stars that remains important until the 17th century

✧ ca. 60 C.E.

The Greek engineer and mathematician Hero of Alexandria creates the aeoliphile, a toylike device that demonstrates the action-reaction principle that is the basis of operation of all rocket engines

✧ ca. 150 C.E.

Greek astronomer Ptolemy writes *Syntaxis* (later called the *Almagest* by Arab astronomers and scholars)—an important book that summarizes all the astronomical knowledge of the ancient astronomers, including the geocentric model of the universe that dominates Western science for more than one and a half millennia

✧ 820

Arab astronomers and mathematicians establish a school of astronomy in Baghdad and translate Ptolemy's work into Arabic, after which it became known as *al-Majisti* (The great work), or the *Almagest,* by medieval scholars

✧ 850

The Chinese begin to use gunpowder for festive fireworks, including a rocketlike device

✧ 1232

The Chinese army uses fire arrows (crude gunpowder rockets on long sticks) to repel Mongol invaders at the battle of Kaifung-fu. This is the first reported use of the rocket in warfare

✧ 1280–90

The Arab historian al-Hasan al-Rammah writes *The Book of Fighting on Horseback and War Strategies,* in which he gives instructions for making both gunpowder and rockets

✧ 1379

Rockets appear in western Europe; they are used in the siege of Chioggia (near Venice), Italy

✧ 1420

The Italian military engineer Joanes de Fontana writes *Book of War Machines,* a speculative work that suggests military applications of gunpowder rockets, including a rocket-propelled battering ram and a rocket-propelled torpedo

✧ 1429

The French army uses gunpowder rockets to defend the city of Orléans. During this period, arsenals throughout Europe begin to test various types of gunpowder rockets as an alternative to early cannons

✧ ca. 1500

According to early rocketry lore, a Chinese official named Wan-Hu attempted to use an innovative rocket-propelled kite assembly to fly through the air. As he sat in the pilot's chair, his servants lit the assembly's 47 gunpowder (black powder) rockets. Unfortunately, this early rocket test pilot disappeared in a bright flash and explosion

✧ 1543

The Polish church official and astronomer Nicolaus Copernicus changes history and initiates the Scientific Revolution with his book *De Revolutionibus Orbium Coelestium* (On the revolutions of the heavenly spheres). This important book, published while Copernicus lay on his deathbed, proposed a Sun-centered (heliocentric) model of the universe in contrast to the longstanding Earth-centered (geocentric) model advocated by Ptolemy and many of the early Greek astronomers

✧ 1608

The Dutch optician Hans Lippershey develops a crude telescope

✧ 1609

The German astronomer Johannes Kepler publishes *New Astronomy*, in which he modifies Nicolaus Copernicus's model of the universe by announcing that the planets have elliptical orbits rather than circular ones. Kepler's laws of planetary motion help put an end to more than 2,000 years of geocentric Greek astronomy

✧ 1610

On January 7, 1610, Galileo Galilei uses his telescope to gaze at Jupiter and discovers the giant planet's four major moons (Callisto, Europa, Io, and Ganymede). He proclaims this and other astronomical observations in his book, *Sidereus Nuncius* (Starry messenger). Discovery of these four Jovian moons encourages Galileo to advocate the heliocentric theory of Nicolaus Copernicus and brings him into direct conflict with church authorities

✧ 1642

Galileo Galilei dies while under house arrest near Florence, Italy, for his clashes with church authorities concerning the heliocentric theory of Nicolaus Copernicus

✧ 1647

The Polish-German astronomer Johannes Hevelius publishes *Seleno-graphia*, in which he provides a detailed description of features on the surface (near side) of the Moon

✧ 1680

Russian czar Peter the Great sets up a facility to manufacture rockets in Moscow. The facility later moves to St. Petersburg and provides the czarist army with a variety of gunpowder rockets for bombardment, signaling, and nocturnal battlefield illumination

✧ 1687

Financed and encouraged by Sir Edmond Halley, Sir Isaac Newton publishes his great work, *Philosophiae Naturalis Principia Mathematica* (Mathematical principles of natural philosophy). This book provides the mathematical foundations for understanding the motion of almost everything in the universe including the orbital motion of planets and the trajectories of rocket-propelled vehicles

✧ 1780s

The Indian ruler Hyder Ally (Ali) of Mysore creates a rocket corps within his army. Hyder's son, Tippo Sultan, successfully uses rockets against the British in a series of battles in India between 1782 and 1799

✧ 1804

Sir William Congreve writes *A Concise Account of the Origin and Progress of the Rocket System* and documents the British military's experience in India. He then starts the development of a series of British military (black-powder) rockets

✧ 1807

The British use about 25,000 of Sir William Congreve's improved military (black-powder) rockets to bombard Copenhagen, Denmark, during the Napoleonic Wars

✧ 1809

The brilliant German mathematician, astronomer, and physicist Carl Friedrich Gauss publishes a major work on celestial mechanics that revolutionizes the calculation of perturbations in planetary orbits. His work paves the way for other 19th-century astronomers to mathematically anticipate and then discover Neptune (in 1846), using perturbations in the orbit of Uranus

✧ 1812

British forces use Sir William Congreve's military rockets against American troops during the War of 1812. British rocket bombardment of Fort William McHenry inspires Francis Scott Key to add "the rocket's red glare" verse in the "Star Spangled Banner"

✧ 1865

The French science fiction writer Jules Verne publishes his famous story *De la terre a la lune* (From the Earth to the Moon). This story interests many people in the concept of space travel, including young readers who go on to become the founders of astronautics: Robert Hutchings Goddard, Hermann J. Oberth, and Konstantin Eduardovich Tsiolkovsky

✧ 1869

American clergyman and writer Edward Everett Hale publishes *The Brick Moon*—a story that is the first fictional account of a human-crewed space station

✧ 1877

While a staff member at the U.S. Naval Observatory in Washington, D.C., the American astronomer Asaph Hall discovers and names the two tiny Martian moons, Deimos and Phobos

✧ 1897

British author H. G. Wells writes the science fiction story *War of the Worlds*—the classic tale about extraterrestrial invaders from Mars

✧ 1903

The Russian technical visionary Konstantin Eduardovich Tsiolkovsky becomes the first person to link the rocket and space travel when he publishes *Exploration of Space with Reactive Devices*

✧ 1918

American physicist Robert Hutchings Goddard writes *The Ultimate Migration*—a far-reaching technology piece within which he postulates the use of an atomic-powered space ark to carry human beings away from a dying Sun. Fearing ridicule, however, Goddard hides the visionary manuscript and it remains unpublished until November 1972—many years after his death in 1945

✧ 1919

American rocket pioneer Robert Hutchings Goddard publishes the Smithsonian monograph *A Method of Reaching Extreme Altitudes*. This important work presents all the fundamental principles of modern rocketry. Unfortunately, members of the press completely miss the true significance of his technical contribution and decide to sensationalize his comments about possibly reaching the Moon with a small, rocket-propelled package. For such "wild fantasy," newspaper reporters dubbed Goddard with the unflattering title of "Moon man"

✧ 1923

Independent of Robert Hutchings Goddard and Konstantin Eduardovich Tsiolkovsky, the German space-travel visionary Hermann J. Oberth publishes the inspiring book *Die Rakete zu den Planetenräumen* (The rocket into planetary space)

✧ 1924

The German engineer Walter Hohmann writes *Die Erreichbarkeit der Himmelskörper* (The attainability of celestial bodies)—an important work that details the mathematical principles of rocket and spacecraft motion. He includes a description of the most efficient (that is, minimum energy) orbit transfer path between two coplanar orbits—a frequently used space operations maneuver now called the Hohmann transfer orbit

✧ 1926

On March 16 in a snow-covered farm field in Auburn, Massachusetts, American physicist Robert Hutchings Goddard makes space technology history by successfully firing the world's first liquid-propellant rocket. Although his primitive gasoline (fuel) and liquid oxygen (oxidizer) device burned for only two and one half seconds and landed about 60 meters away, it represents the technical ancestor of all modern liquid-propellant rocket engines.

In April, the first issue of *Amazing Stories* appears. The publication becomes the world's first magazine dedicated exclusively to science fiction. Through science fact and fiction, the modern rocket and space travel become firmly connected. As a result of this union, the visionary dream for many people in the 1930s (and beyond) becomes that of interplanetary travel

✧ 1929

German space-travel visionary Hermann J. Oberth writes the award-winning book *Wege zur Raumschiffahrt* (Roads to space travel) that helps popularize the notion of space travel among nontechnical audiences

✧ 1933

P. E. Cleator founds the British Interplanetary Society (BIS), which becomes one of the world's most respected space-travel advocacy organizations

✧ 1935

Konstantin Tsiolkovsky publishes his last book, *On the Moon,* in which he strongly advocates the spaceship as the means of lunar and interplanetary travel

✧ 1936

P. E. Cleator, founder of the British Interplanetary Society, writes *Rockets through Space,* the first serious treatment of astronautics in the United Kingdom. However, several established British scientific publications ridicule his book as the premature speculation of an unscientific imagination

✧ 1939–1945

Throughout World War II, nations use rockets and guided missiles of all sizes and shapes in combat. Of these, the most significant with respect to space exploration is the development of the liquid propellant V-2 rocket by the German army at Peenemünde under Wernher von Braun

✧ 1942

On October 3, the German A-4 rocket (later renamed Vengeance Weapon Two or V-2 Rocket) completes its first successful flight from the Peenemünde test site on the Baltic Sea. This is the birth date of the modern military ballistic missile

✧ 1944

In September, the German army begins a ballistic missile offensive by launching hundreds of unstoppable V-2 rockets (each carrying a one-ton high explosive warhead) against London and southern England

✧ 1945

Recognizing the war was lost, the German rocket scientist Wernher von Braun and key members of his staff surrender to American forces near Reutte, Germany in early May. Within months, U.S. intelligence teams, under Operation Paperclip, interrogate German rocket personnel and sort through carloads of captured documents and equipment. Many of these German scientists and engineers join von Braun in the United States to continue their rocket work. Hundreds of captured V-2 rockets are also disassembled and shipped back to the United States.

On May 5, the Soviet army captures the German rocket facility at Peenemünde and hauls away any remaining equipment and personnel. In the closing days of the war in Europe, captured German rocket technology and personnel helps set the stage for the great missile and space race of the cold war

On July 16, the United States explodes the world's first nuclear weapon. The test shot, code named Trinity, occurs in a remote portion of southern New Mexico and changes the face of warfare forever. As part of the cold-war confrontation between the United States and the former Soviet Union, the nuclear-armed ballistic missile will become the most powerful weapon ever developed by the human race.

In October, a then-obscure British engineer and writer, Arthur C. Clarke, suggests the use of satellites at geostationary orbit to support global communications. His article, in *Wireless World* "Extra-Terrestrial Relays," represents the birth of the communications satellite concept—an application of space technology that actively supports the information revolution

✧ 1946

On April 16, the U.S. Army launches the first American-adapted, captured German V-2 rocket from the White Sands Proving Ground in southern New Mexico.

Between July and August the Russian rocket engineer Sergei Korolev develops a stretched-out version of the German V-2 rocket. As part of his engineering improvements, Korolev increases the rocket engine's thrust and lengthens the vehicle's propellant tanks

✧ 1947

On October 30, Russian rocket engineers successfully launch a modified German V-2 rocket from a desert launch site near a place called Kapustin Yar. This rocket impacts about 320 kilometers downrange from the launch site

✧ 1948

The September issue of the *Journal of the British Interplanetary Society* publishes the first in a series of four technical papers by L. R. Shepherd and A. V. Cleaver that explores the feasibility of applying nuclear energy to space travel, including the concepts of nuclear-electric propulsion and the nuclear rocket

✧ 1949

On August 29, the Soviet Union detonates its first nuclear weapon at a secret test site in the Kazakh Desert. Code-named First Lightning (Pervaya Molniya), the successful test breaks the nuclear-weapon monopoly enjoyed by the United States. It plunges the world into a massive nuclear arms race that includes the accelerated development of strategic ballistic missiles capable of traveling thousands of kilometers. Because they are well behind the United States in nuclear weapons technology, the leaders of the former Soviet Union decide to develop powerful, high-thrust rockets to carry their heavier, more primitive-design nuclear weapons. That decision gives the Soviet Union a major launch vehicle advantage when both superpowers decide to race into outer space (starting in 1957) as part of a global demonstration of national power

✧ 1950
On July 24, the United States successfully launches a modified German V-2 rocket with an American-designed WAC Corporal second-stage rocket from the U.S. Air Force's newly established Long Range Proving Ground at Cape Canaveral, Florida. The hybrid, multistage rocket (called Bumper 8) inaugurates the incredible sequence of military missile and space vehicle launches to take place from Cape Canaveral—the world's most famous launch site.

In November, British technical visionary Arthur C. Clarke publishes "Electromagnetic Launching as a Major Contribution to Space-Flight." Clarke's article suggests mining the Moon and launching the mined-lunar material into outer space with an electromagnetic catapult

✧ 1951
Cinema audiences are shocked by the science fiction movie *The Day the Earth Stood Still*. This classic story involves the arrival of a powerful, humanlike extraterrestrial and his robot companion, who come to warn the governments of the world about the foolish nature of their nuclear arms race. It is the first major science fiction story to portray powerful space aliens as friendly, intelligent creatures who come to help Earth.

Dutch-American astronomer Gerard Peter Kuiper suggests the existence of a large population of small, icy planetesimals beyond the orbit of Pluto—a collection of frozen celestial bodies now known as the Kuiper belt

✧ 1952
Collier's magazine helps stimulate a surge of American interest in space travel by publishing a beautifully illustrated series of technical articles written by space experts such as Wernher von Braun and Willey Ley. The first of the famous eight-part series appears on March 22 and is boldly titled "Man Will Conquer Space Soon." The magazine also hires the most influential space artist Chesley Bonestell to provide stunning color illustrations. Subsequent articles in the series introduce millions of American readers to the concept of a space station, a mission to the Moon, and an expedition to Mars

Wernher von Braun publishes *Das Marsprojekt* (The Mars project), the first serious technical study regarding a human-crewed expedition to

Mars. His visionary proposal involves a convoy of 10 spaceships with a total combined crew of 70 astronauts to explore the Red Planet for about one year and then return to Earth

✧ 1953

In August, the Soviet Union detonates its first thermonuclear weapon (a hydrogen bomb). This is a technological feat that intensifies the superpower nuclear arms race and increases emphasis on the emerging role of strategic, nuclear-armed ballistic missiles.

In October, the U.S. Air Force forms a special panel of experts, headed by John von Neumann to evaluate the American strategic ballistic missile program. In 1954, this panel recommends a major reorganization of the American ballistic missile effort

✧ 1954

Following the recommendations of John von Neumann, President Dwight D. Eisenhower gives strategic ballistic missile development the highest national priority. The cold war missile race explodes on the world stage as the fear of a strategic ballistic missile gap sweeps through the American government. Cape Canaveral becomes the famous proving ground for such important ballistic missiles as the Thor, Atlas, Titan, Minuteman, and Polaris. Once developed, many of these powerful military ballistic missiles also serve the United States as space launch vehicles. U.S. Air Force General Bernard Schriever oversees the time-critical development of the Atlas ballistic missile—an astonishing feat of engineering and technical management

✧ 1955

Walt Disney (the American entertainment visionary) promotes space travel by producing an inspiring three-part television series that includes appearances by noted space experts like Wernher von Braun. The first episode, "Man in Space," airs on March 9 and popularizes the dream of space travel for millions of American television viewers. This show, along with its companion episodes, "Man and the Moon" and "Mars and Beyond," make von Braun and the term *rocket scientist* household words

✧ 1957

On October 4, Russian rocket scientist Sergei Korolev with permission from Soviet premier Nikita S. Khrushchev uses a powerful military rocket to successfully place *Sputnik 1* (the world's first artificial satellite) into orbit around Earth. News of the Soviet success sends a political and technical shockwave across the United States. The launch of *Sputnik 1* marks the beginning of the Space Age. It also is the start of the great space race of the cold war—a period when people measure national strength and global prestige by accomplishments (or failures) in outer space.

On November 3, the Soviet Union launches *Sputnik 2*—the world's second artificial satellite. It is a massive spacecraft (for the time) that carries a live dog named Laika, which is euthanized at the end of the mission.

The highly publicized attempt by the United States to launch its first satellite with a newly designed civilian rocket ends in complete disaster on December 6. The Vanguard rocket explodes after rising only a few inches above its launch pad at Cape Canaveral. Soviet successes with *Sputnik 1* and *Sputnik 2* and the dramatic failure of the Vanguard rocket heighten American anxiety. The exploration and use of outer space becomes a highly visible instrument of cold-war politics

✧ 1958

On January 31, the United States successfully launches *Explorer 1*—the first American satellite in orbit around Earth. A hastily formed team from the U.S. Army Ballistic Missile Agency (ABMA) and Caltech's Jet Propulsion Laboratory (JPL), led by Wernher von Braun, accomplishes what amounts to a national prestige rescue mission. The team uses a military ballistic missile as the launch vehicle. With instruments supplied by Dr. James Van Allen of the State University of Iowa, *Explorer 1* discovers Earth's trapped radiation belts—now called the Van Allen radiation belts in his honor.

The National Aeronautics and Space Administration (NASA) becomes the official civilian space agency for the United States government on October 1. On October 7, the newly created NASA announces the start of the Mercury Project—a pioneering program to put the first American astronauts into orbit around Earth.

In mid-December, an entire Atlas rocket lifts off from Cape Canaveral and goes into orbit around Earth. The missile's payload compartment carries Project Score (Signal Communications Orbit Relay Experiment)—a prerecorded Christmas season message from President Dwight D. Eisenhower. This is the first time the human voice is broadcast back to Earth from outer space

✧ 1959

On January 2, the Soviet Union sends a 790 pound-mass (360-kg) spacecraft, *Lunik 1*, toward the Moon. Although it misses hitting the Moon by between 3,125 and 4,375 miles (5,000 and 7,000 km), it is the first human-made object to escape Earth's gravity and go in orbit around the Sun.

In mid-September, the Soviet Union launches *Lunik 2*. The 860 pound-mass (390-kg) spacecraft successfully impacts on the Moon and becomes the first human-made object to (crash-) land on another world. *Lunik 2* carries Soviet emblems and banners to the lunar surface.

On October 4, the Soviet Union sends *Lunik 3* on a mission around the Moon. The spacecraft successfully circumnavigates the Moon and takes the first images of the lunar farside. Because of the synchronous rotation of the Moon around Earth, only the near side of the lunar surface is visible to observers on Earth

✦ 1960

The United States launches the *Pioneer 5* spacecraft on March 11 into orbit around the Sun. The modest-sized (92 pound-mass [42-kg]) spherical American space probe reports conditions in interplanetary space between Earth and Venus over a distance of about 23 million miles (37 million km).

On May 24, the U.S. Air Force launches a MIDAS (Missile Defense Alarm System) satellite from Cape Canaveral. This event inaugurates an important American program of special military surveillance satellites intended to detect enemy missile launches by observing the characteristic infrared (heat) signature of a rocket's exhaust plume. Essentially unknown to the general public for decades because of the classified nature of their mission, the emerging family of missile surveillance satellites provides U.S. government authorities with a reliable early warning system concerning a surprise enemy (Soviet) ICBM attack. Surveillance satellites help support the national policy of strategic nuclear deterrence throughout the cold war and prevent an accidental nuclear conflict.

The U.S. Air Force successfully launches the *Discoverer 13* spacecraft from Vandenberg Air Force Base on August 10. This spacecraft is actually part of a highly classified Air Force and Central Intelligence Agency (CIA) reconnaissance satellite program called Corona. Started under special executive order from President Dwight D. Eisenhower, the joint agency spy satellite program begins to provide important photographic images of denied areas of the world from outer space. On August 18, *Discoverer 14* (also called *Corona XIV*) provides the U.S. intelligence community its first satellite-acquired images of the former Soviet Union. The era of satellite reconnaissance is born. Data collected by the spy satellites of the National Reconnaissance Office (NRO) contribute significantly to U.S. national security and help preserve global stability during many politically troubled times.

On August 12, NASA successfully launches the *Echo 1* experimental spacecraft. This large (100 foot [30.5 m] in diameter) inflatable, metalized balloon becomes the world's first passive communications satellite. At the dawn of space-based telecommunications, engineers bounce radio signals off the large inflated satellite between the United States and the United Kingdom.

The former Soviet Union launches *Sputnik 5* into orbit around Earth. This large spacecraft is actually a test vehicle for the new *Vostok* spacecraft that will soon carry cosmonauts into outer space. *Sputnik 5* carries two dogs, Strelka and Belka. When the spacecraft's recovery capsule functions properly the next day, these two dogs become the first living creatures to return to Earth successfully from an orbital flight

✦ 1961

On January 31, NASA launches a Redstone rocket with a Mercury Project space capsule on a suborbital flight from Cape Canaveral. The passenger astrochimp Ham is safely recovered down range in the Atlantic Ocean after reaching an altitude of 155 miles (250 km). This successful primate space mission is a key step in sending American astronauts safely into outer space.

The Soviet Union achieves a major space exploration milestone by successfully launching the first human being into orbit around Earth. Cosmonaut Yuri Gagarin travels into outer space in the *Vostok 1* spacecraft and becomes the first person to observe Earth directly from an orbiting space vehicle.

On May 5, NASA uses a Redstone rocket to send astronaut Alan B. Shepard, Jr., on his historic 15-minute suborbital flight into outer space from Cape Canaveral. Riding inside the Mercury Project *Freedom 7* space capsule, Shepard reaches an altitude of 115 miles (186 km) and becomes the first American to travel in space.

President John F. Kennedy addresses a joint session of the U.S. Congress on May 25. In an inspiring speech touching on many urgent national needs, the newly elected president creates a major space challenge for the United States when he declares: "I believe that this nation should commit itself to achieving the goal, before this decade is out, of landing a man on the Moon and returning him safely to Earth." Because of his visionary leadership, when American astronauts Neil A. Armstrong and Edwin E. "Buzz" Aldrin, Jr., step onto the lunar surface for the first time on July 20, 1969, the United States is recognized around the world as the undisputed winner of the cold-war space race

✦ 1962

On February 20, astronaut John Herschel Glenn, Jr., becomes the first American to orbit Earth in a spacecraft. An Atlas rocket launches the NASA Mercury Project *Friendship 7* space capsule from Cape Canaveral. After completing three orbits, Glenn's capsule safely splashes down in the Atlantic Ocean.

In late August, NASA sends the *Mariner 2* spacecraft to Venus from Cape Canaveral. *Mariner 2* passes within 21,700 miles (35,000 km) of the planet on December 14, 1962—thereby becoming the world's first successful interplanetary space probe. The spacecraft observes very high surface temperatures (~800°F [430°C]). These data shatter pre–space age visions about Venus being a lush, tropical planetary twin of Earth.

During October, the placement of nuclear-armed Soviet offensive ballistic missiles in Fidel Castro's Cuba precipitates the Cuban Missile Crisis. This dangerous superpower confrontation brings the world perilously close to nuclear warfare. Fortunately, the crisis dissolves when Premier Nikita S. Khrushchev withdraws the Soviet ballistic missiles after much skillful political maneuvering by President John F. Kennedy and his national security advisers

✧ 1964

On November 28, NASA's *Mariner 4* spacecraft departs Cape Canaveral on its historic journey as the first spacecraft from Earth to visit Mars. It successfully encounters the Red Planet on July 14, 1965 at a flyby distance of about 6,100 miles (9,800 km). *Mariner 4*'s closeup images reveal a barren, desertlike world and quickly dispel any pre–space age notions about the existence of ancient Martian cities or a giant network of artificial canals

✧ 1965

A Titan II rocket carries astronauts Virgil "Gus" I. Grissom and John W. Young into orbit on March 23 from Cape Canaveral, inside a two-person Gemini Project spacecraft. NASA's *Gemini 3* flight is the first crewed mission for the new spacecraft and marks the beginning of more sophisticated space activities by American crews in preparation for the Apollo Project lunar missions

✧ 1966

The former Soviet Union sends the *Luna 9* spacecraft to the Moon on January 31. The 220 pound-mass (100-kg) spherical spacecraft soft lands in the Ocean of Storms region on February 3, rolls to a stop, opens four petal-like covers, and then transmits the first panoramic television images from the Moon's surface.

The former Soviet Union launches the *Luna 10* to the Moon on March 31. This massive (3,300 pound-mass [1,500-kg]) spacecraft becomes the first human-made object to achieve orbit around the Moon.

On May 30, NASA sends the *Surveyor 1* lander spacecraft to the Moon. The versatile robot spacecraft successfully makes a soft landing (June 1) in the Ocean of Storms. It then transmits over 10,000 images from the lunar surface and performs numerous soil mechanics experiments in preparation for the Apollo Project human landing missions.

In mid-August, NASA sends the *Lunar Orbiter 1* spacecraft to the Moon from Cape Canaveral. It is the first of five successful missions to collect detailed images of the Moon from lunar orbit. At the end of each mapping mission the orbiter spacecraft is intentionally crashed into the Moon to prevent interference with future orbital activities

✧ 1967

On January 27, disaster strikes NASA's Apollo Project. While inside their *Apollo 1* spacecraft during a training exercise on Launch Pad 34 at Cape Canaveral, astronauts Virgil "Gus" I. Grissom, Edward H. White, Jr., and Roger B. Chaffee are killed when a flash fire sweeps through their space-craft. The Moon landing program was delayed by 18 months, while major design and safety changes are made in the Apollo Project spacecraft.

On April 23, tragedy also strikes the Russian space program when the Soviets launch cosmonaut Vladimir Komarov in the new *Soyuz* (union) spacecraft. Following an orbital mission plagued with difficulties, Koma-rov dies (on April 24) during reentry operations, when the spacecraft's parachute fails to deploy properly and the vehicle hits the ground at high speed

✧ 1968

On December 21, NASA's *Apollo 8* spacecraft (command and service modules only) departs Launch Complex 39 at the Kennedy Space Center during the first flight of mighty Saturn V launch vehicle with a human crew as part of the payload. Astronauts Frank Borman, James Arthur Lovell, Jr., and William A. Anders become the first people to leave Earth's gravitational influence. They go into orbit around the Moon and capture images of an incredibly beautiful Earth "rising" above the starkly barren lunar horizon—pictures that inspire millions and stimulate an emerging environmental movement. After 10 orbits around the Moon, the first lunar astronauts return safely to Earth on December 27

✧ 1969

The entire world watches as NASA's *Apollo 11* mission leaves for the Moon on July 16 from the Kennedy Space Center. Astronauts Neil A. Armstrong, Michael Collins, and Edwin E. "Buzz" Aldrin, Jr., make a long-held dream of humanity a reality. On July 20, American astronaut Neil Armstrong cau-tiously descends the steps of the lunar excursion module's ladder and steps on the lunar surface, stating, "One small step for a man, one giant leap for mankind!" He and Buzz Aldrin become the first two people to walk on another world. Many people regard the Apollo Project lunar landings as the greatest technical accomplishment in all of human history

✧ 1970

NASA's *Apollo 13* mission leaves for the Moon on April 11. Suddenly, on April 13, a life-threatening explosion occurs in the service module portion of the Apollo spacecraft. Astronauts James A. Lovell, Jr., John Leonard Swigert, and Fred Wallace Haise, Jr., must use their lunar excursion module (LEM) as a lifeboat. While an anxious world waits and listens, the crew skillfully maneuvers their disabled spacecraft around the Moon. With critical supplies running low, they limp back to Earth on a free-return trajectory. At just the right moment on April 17, they abandon the LEM *Aquarius* and board the Apollo Project spacecraft (command module) for a successful atmospheric reentry and recovery in the Pacific Ocean

✧ 1971

On April 19, the former Soviet Union launches the first space station (called *Salyut 1*). It remains initially uncrewed because the three-cosmonaut crew of the *Soyuz 10* mission (launched on April 22) attempts to dock with the station but cannot go on board

✧ 1972

In early January, President Richard M. Nixon approves NASA's space shuttle program. This decision shapes the major portion of NASA's program for the next three decades.

On March 2, an Atlas-Centaur launch vehicle successfully sends NASA's *Pioneer 10* spacecraft from Cape Canaveral on its historic mission. This far-traveling robot spacecraft becomes the first to transit the main-belt asteroids, the first to encounter Jupiter (December 3, 1973) and by crossing the orbit of Neptune on June 13, 1983 (which at the time was the farthest planet from the Sun) the first human-made object ever to leave the planetary boundaries of the solar system. On an interstellar trajectory, *Pioneer 10* (and its twin, *Pioneer 11*) carries a special plaque, greeting any intelligent alien civilization that might find it drifting through interstellar space millions of years from now.

On December 7, NASA's *Apollo 17* mission, the last expedition to the Moon in the 20th century, departs from the Kennedy Space Center, propelled by a mighty Saturn V rocket. While astronaut Ronald E. Evans remains in lunar orbit, fellow astronauts Eugene A. Cernan and Harrison H. Schmitt become the 11th and 12th members of the exclusive Moon walkers club. Using a lunar rover, they explore the Taurus-Littrow region. Their safe return to Earth on December 19 brings to a close one of the epic periods of human exploration

✧ 1973

In early April, while propelled by Atlas-Centaur rocket, NASA's *Pioneer 11* spacecraft departs on an interplanetary journey from Cape Canaveral. The spacecraft encounters Jupiter (December 2, 1974) and then uses a gravity assist maneuver to establish a flyby trajectory to Saturn. It is the first spacecraft to view Saturn at close range (closest encounter on September 1, 1979) and then follows a path into interstellar space.

On May 14, NASA launches *Skylab*—the first American space station. A giant Saturn V rocket is used to place the entire large facility into orbit in a single launch. The first crew of three American astronauts arrives on May 25 and makes the emergency repairs necessary to save the station, which suffered damage during the launch ascent. Astronauts Charles (Pete) Conrad, Jr., Paul J. Weitz, and Joseph P. Kerwin stay onboard for 28 days. They are replaced by astronauts Alan L. Bean, Jack R. Lousma, and Owen K. Garriott, who arrive on July 28 and live in space for about 59 days. The final *Skylab* crew (astronauts Gerald P. Carr, William R. Pogue, and Edward G. Gibson) arrive on November 11 and resided in the station until February 8, 1974—setting a space endurance record (for the time) of 84 days. NASA then abandons *Skylab*.

In early November, NASA launches the *Mariner 10* spacecraft from Cape Canaveral. It encounters Venus (February 5, 1974) and uses a gravity assist maneuver to become the first spacecraft to investigate Mercury at close range

✧ 1975

In late August and early September, NASA launches the twin *Viking 1* (August 20) and *Viking 2* (September 9) orbiter/lander combination spacecraft to the Red Planet from Cape Canaveral. Arriving at Mars in 1976, all Viking Project spacecraft (two landers and two orbiters) perform exceptionally well—but the detailed search for microscopic alien life-forms on Mars remains inconclusive

✧ 1977

On August 20, NASA sends the *Voyager 2* spacecraft from Cape Canaveral on an epic grand tour mission during which it encounters all four giant planets and then departs the solar system on an interstellar trajectory. Using the gravity assist maneuver, *Voyager 2* visits Jupiter (July 9, 1979), Saturn (August 25, 1981), Uranus (January 24, 1986), and Neptune (August 25, 1989). The resilient, far-traveling robot spacecraft (and its twin *Voyager 1*) also carries a special interstellar message from Earth—a digital record entitled *The Sounds of Earth*.

On September 5, NASA sends the *Voyager 1* spacecraft from Cape Canaveral on its fast trajectory journey to Jupiter (March 5, 1979), Saturn (March 12, 1980), and beyond the solar system

✧ 1978

In May, the British Interplanetary Society releases its Project Daedalus report—a conceptual study about a one-way robot spacecraft mission to Barnard's star at the end of the 21st century

✧ 1979

On December 24, the European Space Agency successfully launches the first Ariane 1 rocket from the Guiana Space Center in Kourou, French Guiana

✧ 1980

India's Space Research Organization successfully places a modest 77 pound-mass (35 kg) test satellite (called *Rohini*) into low Earth orbit on July 1. The launch vehicle is a four-stage, solid propellant rocket manufactured in India. The SLV-3 (Standard Launch Vehicle-3) gives India independent national access to outer space

✧ 1981

On April 12, NASA launches the space shuttle *Columbia* on its maiden orbital flight from Complex 39-A at the Kennedy Space Center. Astronauts John W. Young and Robert L. Crippen thoroughly test the new aerospace vehicle. Upon reentry, it becomes the first spacecraft to return to Earth by gliding through the atmosphere and landing like an airplane. Unlike all previous onetime use space vehicles, *Columbia* is prepared for another mission in outer space

✧ 1986

On January 24, NASA's *Voyager 2* spacecraft encounters Uranus.

On January 28, the space shuttle *Challenger* lifts off from the NASA Kennedy Space Center on its final voyage. At just under 74 seconds into the STS 51-L mission, a deadly explosion occurs, killing the crew and destroying the vehicle. Led by President Ronald Reagan, the United States mourns seven astronauts lost in the *Challenger* accident

✧ 1988

On September 19, the State of Israel uses a Shavit (comet) three-stage rocket to place the country's first satellite (called *Ofeq 1*) into an unusual east-to-west orbit—one that is opposite to the direction of Earth's rotation but necessary because of launch safety restrictions.

As the *Discovery* successfully lifts off on September 29 for the STS-26 mission, NASA returns the space shuttle to service following a 32-month hiatus after the *Challenger* accident

✧ 1989

On August 25, the *Voyager 2* spacecraft encounters Neptune

✧ 1994

In late January, a joint Department of Defense and NASA advanced technology demonstration spacecraft, *Clementine,* lifts off for the Moon from Vandenberg Air Force Base. Some of the spacecraft's data suggest that the Moon may actually possess significant quantities of water ice in its permanently shadowed polar regions

✧ 1995

In February, during NASA's STS-63 mission, the space shuttle *Discovery* approaches (encounters) the Russian *Mir* space station as a prelude to the development of the *International Space Station.* Astronaut Eileen Marie Collins serves as the first female shuttle pilot.

On March 14, the Russians launch the *Soyuz TM-21* spacecraft to the *Mir* space station from the Baikanour Cosmodrome. The crew of three includes American astronaut Norman Thagard—the first American to travel into outer space on a Russian rocket and the first to stay on the *Mir* space station. The *Soyuz TM-21* cosmonauts also relieve the previous *Mir* crew, including cosmonaut Valeri Polyakov, who returns to Earth on March 22 after setting a world record for remaining in space for 438 days.

In late June, NASA's space shuttle *Atlantis* docks with the Russian *Mir* space station for the first time. During this shuttle mission (STS-71), *Atlantis* delivers the *Mir 19* crew (cosmonauts Anatoly Solovyev and Nikolai Budarin) to the Russian space station and then returns the *Mir 18* crew back to Earth—including American astronaut Norman Thagard, who has just spent 115 days in space onboard the *Mir.* The Shuttle-*Mir* docking program is the first phase of the *International Space Station.* A total of nine shuttle-*Mir* docking missions will occur between 1995 and 1998

✧ 1998

In early January, NASA sends the *Lunar Prospector* to the Moon from Cape Canaveral. Data from this orbiter spacecraft reinforces previous hints that the Moon's polar regions may contain large reserves of water ice in a mixture of frozen dust lying at the frigid bottom of some permanently shadowed craters.

In early December, the space shuttle *Endeavour* ascends from the NASA Kennedy Space Center on the first assembly mission of the *International Space Station*. During the STS-88 shuttle mission, *Endeavour* performs a rendezvous with the previously launched Russian-built *Zarya* (sunrise) module. An international crew connects this module with the American-built *Unity* module carried in the shuttle's cargo bay

✧ 1999

In July, astronaut Eileen Marie Collins serves as the first female space shuttle commander (STS-93 mission) as the *Columbia* carries NASA's *Chandra X-ray Observatory* into orbit

✧ 2001

NASA launches the *Mars Odyssey 2001* mission to the Red Planet in early April—the spacecraft successfully orbits the planet in October

✧ 2002

On May 4, NASA successfully launches its *Aqua* satellite from Vandenberg Air Force Base. This sophisticated Earth-observing spacecraft joins the *Terra* spacecraft in performing Earth system science studies.

On October 1, the United States Department of Defense forms the U.S. Strategic Command (USSTRATCOM) as the control center for all American strategic (nuclear) forces. USSTRATCOM also conducts military space operations, strategic warning and intelligence assessment, and global strategic planning

✧ 2003

On February 1, while gliding back to Earth after a successful 16-day scientific research mission (STS-107), the space shuttle *Columbia* experiences a catastrophic reentry accident at an altitude of about 63 km over the Western United States. Traveling at 18 times the speed of sound, the orbiter vehicle disintegrates, taking the lives of all seven crew members: six American astronauts (Rick Husband, William McCool, Michael Anderson, Kalpana Chawla, Laurel Clark, and David Brown) and the first Israeli astronaut (Ilan Ramon).

NASA's Mars Exploration Rover (MER) *Spirit* is launched by a Delta II rocket to the Red Planet on June 10. *Spirit*, also known as MER-A, arrives safely on Mars on January 3, 2004 and begins its teleoperated surface exploration mission under the supervision of mission controllers at the NASA Jet Propulsion Laboratory.

NASA launches the second Mars Exploration Rover, called *Opportunity,* using a Delta II rocket launch, which lifts off from Cape Canaveral Air Force Station on July 7, 2003. *Opportunity,* also called MER-B, successfully lands on Mars on January 24, 2004, and starts its teleoperated surface exploration mission under the supervision of mission controllers at the NASA Jet Propulsion Laboratory

✧ 2004

On July 1, NASA's *Cassini* spacecraft arrives at Saturn and begins its four-year mission of detailed scientific investigation.

In mid-October, the Expedition 10 crew, riding a Russian launch vehicle from Baikonur Cosmodrome, arrives at the *International Space Station* and the Expedition 9 crew returns safely to Earth.

On December 24, the 703 pound-mass (319-kg) *Huygens* probe successfully separates from the *Cassini* spacecraft and begins its journey to Saturn's moon, Titan

✧ 2005

On January 14, the *Huygens* probe enters the atmosphere of Titan and successfully reaches the surface some 147 minutes later. *Huygens* is the first spacecraft to land on a moon in the outer solar system.

On July 4, NASA's Deep Impact mission successfully encountered Comet Tempel 1.

NASA successfully launched the space shuttle *Discovery* on the STS-114 mission on July 26 from the Kennedy Space Center in Florida. After docking with the *International Space Station*, the *Discovery* returned to Earth and landed at Edwards AFB, California, on August 9.

On August 12, NASA launched the *Mars Reconnaissance Orbiter* from Cape Canaveral AFS, Florida.

On September 19, NASA announced plans for a new spacecraft designed to carry four astronauts to the Moon and to deliver crews and supplies to the *International Space Station*. NASA also introduced two new, shuttle-derived launch vehicles: a crew-carrying rocket and a cargo-carrying, heavy-lift rocket.

The Expedition 12 crew (Commander William McArthur and Flight Engineer Valery Tokarev) arrived at the *International Space Station* on October 3 and replaced the Expedition 11 crew.

The People's Republic of China successfully launched its second human spaceflight mission, called *Shenzhou 6*, on October 12. Two tai-konauts, Fei Junlong and Nie Haisheng, traveled in space for almost five days and made 76 orbits of Earth before returning safely to Earth, making a soft, parachute-assisted landing in northern Inner Mongolia

✧ 2006

On January 15, the sample package from NASA's *Stardust* spacecraft, containing comet samples, successfully returned to Earth.

NASA launched the *New Horizons* spacecraft from Cape Canaveral on January 19 and successfully sent this robot probe on its long one-way mission to conduct a scientific encounter with the Pluto system (in 2015) and then to explore portions of the Kuiper belt that lie beyond.

Follow-up observations by NASA's *Hubble Space Telescope*, reported on February 22, have confirmed the presence of two new moons around the distant planet Pluto. The moons, tentatively called S/2005 P 1 and S/2005 P 2, were first discovered by *Hubble* in May 2005, but the science team wanted to further examine the Pluto system to characterize the orbits of the new moons and validate the discovery.

NASA scientists announced on March 9 that the *Cassini* spacecraft may have found evidence of liquid water reservoirs that erupt in Yellowstone Park–like geysers on Saturn's moon Enceladus.

On March 10, NASA's *Mars Reconnaissance Orbiter* successfully arrived at Mars and began a six-month-long process of adjusting and trimming the shape of its orbit around the Red Planet prior to performing its operational mapping mission.

The Expedition 13 crew (Commander Pavel Vinogradov and Flight Engineer Jeff Williams) arrived at the *International Space Station* on April 1 and replaced the Expedition 12 crew. Joining them for several days before returning back to Earth with the Expedition 12 crew was Brazil's first astronaut, Marcos Pontes

Glossary

abort To cut short or cancel an operation with a rocket, spacecraft, or aerospace vehicle, especially because of equipment failure. NASA's space shuttle system has two types of abort modes during the ascent phase of a flight: intact abort and contingency abort. An intact abort is designed to achieve a safe return of the astronaut crew and orbiter vehicle to a planned landing site. A contingency abort involves a ditching operation in which the crew is saved, but the orbiter vehicle is damaged or destroyed.

acceleration (symbol: a) The rate at which the velocity (speed) of an object changes with time. Acceleration is a vector quantity and has the physical dimensions of length per unit time to the second power (for example, meters per second per second $[m/s^2]$).

acceleration of gravity The local acceleration due to gravity on or near the surface of a planet. On Earth, the acceleration due to gravity (symbol: g) of a free-falling object has the standard value of 32.1740 feet per second per second ($9.80665 m/s^2$) by international agreement.

acronym A word formed from the first letters of a name, such as *HST*, which means the *Hubble Space Telescope*, or a word formed by combining the initial parts of a series of words, such as *lidar*, which means light detection and ranging. Acronyms are frequently used in the naming of space technology.

aerospace A term, derived from *aeronautics* and *space*, meaning of or pertaining to Earth's atmospheric envelope and outer space beyond it. NASA's space shuttle Orbiter vehicle is called an aerospace vehicle, because it operates both in the atmosphere and in outer space.

aerospace ground equipment (AGE) All the support and test equipment needed on Earth's surface to make an aerospace system or spacecraft function properly during its intended space mission.

aerozine   A liquid rocket fuel consisting of a mixture of hydrazine (N_2H_4) and unsymmetrical dimethylhydrazine (acronym: UDMH). Its chemical formula is $(CH_3)_2NNH_2$.

Agena   A versatile upper stage rocket that supported numerous American military and civilian space missions in the 1960s and 1970s. One special feature of this liquid propellant system was its in-space engine restart capability.

air launch   The process of launching a guided missile or rocket from an aircraft while it is in flight.

antimatter   Matter in which the ordinary nuclear particles (such as electrons, protons, and neutrons) are replaced by their corresponding antiparticles—that is, positrons, antiprotons, antineutrons, and so on. Sometimes called *mirror matter*. Normal matter and antimatter mutually annihilate each other upon contact and are converted into pure energy, called annihilation radiation.

antislosh baffle   A device installed in the propellant tank of a liquid-fuel rocket to dampen unwanted liquid motion or sloshing during flight.

Apollo Project   The American effort in the 1960s and early 1970s to successfully place astronauts on the surface of the Moon and return them safely to Earth. The project was launched in May 1961 by President John F. Kennedy in response to a growing space technology challenge from the Soviet Union. Managed by NASA, the *Apollo 8* mission sent the first three humans to the vicinity of the Moon in December 1968. The *Apollo 11* mission involved the first human landing on another world (20 July 1969). *Apollo 17,* the last lunar landing mission under this project, took place in December 1972. The project is often considered one of the greatest technical accomplishments in all of human history.

arc-jet engine   An electric rocket engine that heats a propellant gas by passing through it an electric arc.

Ariane   Family of modern launch vehicles developed by the French Space Agency (Centre National d'Etudes Spatiales, or CNES) and the European Space Agency (ESA). The Ariane 4 rocket, Europe's "space workhorse," and the newer, more powerful Ariane 5 rocket have carried many scientific and commercial payloads into orbit from the Guiana Space Center in Kourou, French Guiana.

astro- A prefix that means "star" or (by extension) outer space or celestial; for example, astronaut, astronautics, or astrophysics.

astrodynamics The application of celestial mechanics, propulsion system theory, and related fields of science and engineering to the problem of carefully planning and directing the trajectory of a space vehicle.

astronautics The branch of engineering science dealing with spaceflight and the design and operation of space vehicles.

backout The process of undoing tasks that have already been completed during the countdown of a launch vehicle, usually in reverse order.

Baikonur Cosmodrome The major launch site for the space program of the former Soviet Union and later the Russian Federation. The complex is located just east of the Aral Sea in Kazakhstan (now an independent republic); also known as the Tyuratam launch site during the cold war. The Soviets launched *Sputnik 1* (1957), the first artificial satellite, and cosmonaut Yuri Gagarin, the first human to fly in outer space (1961), from this location.

ballistic missile A missile that is propelled by rocket engines and guided only during the initial (thrust-producing) phase of its flight. In the non-powered and nonguided phase of its flight, it assumes a ballistic trajectory similar to that of an artillery shell. After thrust termination, reentry vehicles (RVs) can be released and these RVs also follow free-falling (ballistic) trajectories toward their targets. *Compare with* GUIDED MISSILE.

ballistic missile defense (BMD) A proposed defense system designed to protect a territory from incoming ballistic missiles and their warhead-carrying reentry vehicles. A variety of BMD technologies have been suggested, including high-energy laser weapons, high performance interceptor missiles, and kinetic energy weapon systems. However, the BMD problem is technically challenging and can be likened to stopping an incoming high-velocity rifle bullet with another rifle bullet.

ballistic trajectory The path an object (that does not have lifting surfaces) follows while being acted upon by only the force of gravity and any resistive aerodynamic forces of the medium through which it passes. A stone tossed into the air follows a ballistic trajectory. Similarly, after its propulsive unit stops operating, a rocket vehicle describes a ballistic trajectory.

beam rider   A missile guided to its target by a beam of electromagnetic radiation, such as a radar beam or a laser beam.

bell nozzle   A nozzle with a circular opening for a throat and an axisymmetric contoured wall downstream of the throat that gives this type of nozzle a characteristic bell shape.

bipropellant rocket   A rocket that uses two unmixed (uncombined) liquid chemicals as its fuel and oxidizer, respectively. The two chemical propellants flow separately into the rocket's combustion chamber, where they are combined and combusted to produce high-temperature, thrust-generating gases. The combustion gases then exit the rocket system through a suitably designed nozzle.

bird   A popular aerospace industry expression (jargon) for a rocket, missile, satellite, or spacecraft.

blastoff   The moment a rocket or aerospace vehicle rises from its launch pad under full thrust. *See also* LIFTOFF.

blockhouse (or block house)   A reinforced-concrete structure, often built partially underground, that provides protection against blast, heat, and possibly an abort explosion during rocket launchings.

boiloff   The loss of a cryogenic propellant, such as liquid oxygen or liquid hydrogen, due to vaporization. This happens when the temperature of the cryogenic propellant rises slightly in the propellant tank of a rocket being prepared for launch. The longer a fully-fueled rocket vehicle sits on its launch pad, the more significant the problem of boiloff becomes.

booster rocket   A rocket motor, either solid or liquid propellant, that assists the main propulsive system (called the sustainer engine) of a launch vehicle during some part of its flight.

British thermal unit (Btu)   A quantity of energy, especially thermal energy (heat). Defined as the amount of heat needed to increase the temperature of 1 pound-mass (0.45 kg) of water 1°F (0.56°C) at normal atmospheric pressure.

burnout   The moment in time or the point in a rocket's trajectory when combustion of fuels in the engine is terminated. This usually occurs when all the propellants are consumed.

bus The rocket-propelled final stage of an intercontinental ballistic missile that, after booster burnout, places warheads and (possibly) decoys on ballistic trajectories toward their targets. Also called the post-boost vehicle.

canard A horizontal trim and control surface on an aerodynamic vehicle.

cannibalizing The process of taking functioning parts from a non-operating spacecraft or launch vehicle and installing these salvaged parts in another spacecraft or launch vehicle in order to make the latter operational.

Cape Canaveral The region on Florida's east central coast from which the United States Air Force and NASA have launched more than 3,000 rockets since 1950. Cape Canaveral Air Force Station is the major east coast launch site for the Department of Defense, while the adjacent NASA Kennedy Space Center is the spaceport for the fleet of space shuttle vehicles.

captive firing The firing of a rocket propulsion system at full or partial thrust while the rocket is restrained in a test stand facility. Usually engineers instrument the propulsion system to obtain test data that verify rocket design and demonstrate performance levels. Sometimes called a holddown test.

cavitation The formation of bubbles (vapor-filled cavities) in a flowing liquid. The formation of these cavities can adversely impact the performance of high-speed hydraulic machinery, such as the turbopump system of a liquid propellant rocket engine.

Centaur (rocket) A powerful and versatile upper-stage rocket originally developed by the United States in the 1950s for use with the Atlas launch vehicle. Engineered by Krafft Ehricke, it was the first American rocket to use liquid hydrogen as its propellant. Centaur has supported many important military and scientific missions, including the Cassini mission (launched on October 15, 1997) to Saturn.

Centre National d'Etudes Spatiales (CNES) The public body responsible for all aspects of French space activity including launch vehicles and spacecraft. CNES has four main centers: Headquarters (Paris), the Launch Division at Evry, the Toulouse Space Center, and the Guiana Space Center (launch site) in Kourou, French Guiana (South America). *See also* ARIANE and EUROPEAN SPACE AGENCY.

Challenger **accident**   NASA's space shuttle *Challenger* took off from Complex 39-B at the Kennedy Space Center on January 28, 1986 as part of the STS 51-L mission. Seventy-four seconds into the flight, an explosion occurred that caused the loss of the aerospace vehicle and its entire crew, including astronauts Francis Scobee, Michael Smith, Ellison Onizuka, Judith Resnik, Ronald McNair, Christa McAuliffe, and Gregory Jarvis.

chemical rocket   A rocket that uses the combustion of a chemical fuel in either solid or liquid form to generate thrust. The chemical fuel requires an oxidizer to support combustion.

chilldown   Cooling all or part of a cryogenic (very cold) rocket engine system from ambient (room) temperature down to cryogenic temperature by circulating cryogenic propellant (fluid) through the system prior to engine start.

choked flow   A flow condition in a duct or pipe such that the flow upstream of a certain critical section (like a nozzle or valve) cannot be increased by further reducing downstream pressure.

cold-flow test   The thorough testing of a liquid propellant rocket engine without actually firing (igniting) it. This type test helps aerospace engineers verify the performance and efficiency of a propulsion system, since all aspects of propellant flow and conditioning, except combustion, are examined. Tank pressurization, propellant loading, and propellant flow into the combustion chamber (without ignition) are usually included in a cold-flow test.

cold war   The ideological conflict between the United States and the former Soviet Union from approximately 1946 to 1989, involving rivalry, mistrust, and hostility just short of overt military action. The tearing down of the Berlin Wall in November 1989 generally is considered as the (symbolic) end of the cold-war period.

Columbia **accident**   While gliding back to Earth on February 1, 2003, after a very successful 16-day scientific research mission in low Earth orbit, NASA's space shuttle *Columbia* experienced a catastrophic reentry accident and broke apart at an altitude of about 63 kilometers over Texas. Traveling through the upper atmosphere at approximately 18 times the speed of sound, the Orbiter vehicle disintegrated, taking the lives of its seven crewmembers: six American astronauts (Rick D. Husband, William C. McCool, Michael P. Anderson, Kalpana Chawla, Laurel Blair Salton Clark, and David M. Brown) and the first Israeli astronaut (Ilan Ramon). Post-accident investigations indicate that a severe heating problem occurred in *Columbia*'s left wing, as a result of structural damage from debris impact during launch.

combustion chamber   The part of a rocket engine in which the combustion of chemical propellants takes place at high pressure. The combustion chamber and the diverging section of the nozzle make up a rocket's thrust chamber. Sometimes called the firing chamber or simply the chamber.

command destruct   An intentional action leading to the destruction of a rocket or missile in flight. Whenever a malfunctioning vehicle's performance creates a safety hazard on or off the rocket test range, the range safety officer sends the command-destruct signal to destroy it.

companion body   A nose cone, protective shroud, last-stage rocket, or payload separation hardware that orbits Earth along with an operational satellite or spacecraft. Companion bodies contribute significantly to a growing space (orbital) debris population in low Earth orbit.

control rocket   A low-thrust rocket, such as a retrorocket or a vernier engine, used to guide, to change the attitude of, or to make small corrections in the velocity of an aerospace vehicle, spacecraft, or expendable launch vehicle.

converging-diverging (CD) nozzle   A thrust-producing flow device for expanding and accelerating hot exhaust gases from a rocket engine. A properly designed nozzle efficiently converts the thermal energy of combustion into kinetic energy of the combustion product gases. In a supersonic converging-diverging nozzle, the hot gas upstream of the nozzle throat is at subsonic velocity (i.e., the Mach Number (M) < 1), reaches sonic velocity (the speed of sound, for which M = 1) at the throat of the nozzle, and then expands to supersonic velocity (M > 1) downstream of the nozzle throat region while flowing through the diverging section of the nozzle. *See also* DE LAVAL NOZZLE.

countdown   The step-by-step process that leads to the launch of a rocket or aerospace vehicle. A countdown takes place in accordance with a specific schedule—with zero being the "go" or activate time.

cryogenic propellant   A rocket fuel, oxidizer, or propulsion fluid that is liquid only at very low (cryogenic) temperatures. Liquid hydrogen (LH_2) and liquid oxygen (LO_2) are examples.

cutoff   The act of shutting off the propellant flow in a rocket or of stopping the combustion of the propellant. *Compare with* BURNOUT.

deboost   A retrograde (opposite-direction) burn of one or more low-thrust rockets or an aerobraking maneuver that lowers the altitude of an orbiting spacecraft.

De Laval nozzle   A flow device that efficiently converts the energy content of a hot, high-pressure gas into the kinetic energy. Originally developed by Carl de Laval for use in certain steam turbines, this versatile converging-diverging nozzle is now used in practically all modern rockets. The device constricts the outflow of the high pressure (combustion) gas until it reaches the velocity of sound (at the nozzle's throat) and then expands the exiting gas to very high velocities.

Delta (launch vehicle)   A versatile family of American two- and three-stage liquid propellant, expendable launch vehicles that use multiple strap-on booster rockets in several configurations. The Delta rocket vehicle family has successfully launched more than 250 U.S. and foreign satellites, earning it the nickname space workhorse vehicle.

delta-V (symbol: ΔV)   In aerospace engineering, a velocity change; a useful numerical index of the maneuverability of a spacecraft or rocket. This term often represents the maximum change in velocity that a space vehicle's propulsion system can provide. For example, the delta-V capability of an upper-stage propulsion system used to move a satellite from a lower altitude orbit to a higher altitude orbit or to place an Earth-orbiting spacecraft on an interplanetary trajectory. Often described in terms of feet per second (m/s).

destruct   In aerospace operations, the deliberate destruction of a missile or rocket vehicle after it has been launched but before it has completed its course. Destruct commands are executed by the range safety officer, whenever a missile or rocket veers off its intended (plotted) course or functions in a way so as to become a hazard to life or property. *See also* COMMAND DESTRUCT.

destruct line   A boundary line on a rocket test range that lies on each side of the downrange course. For safety reasons, a rocket or missile is not allowed to fly across this line. If the vehicle's flight path touches the destruct line, it is destroyed by the range safety officer, who enforces established command-destruct procedures. The impact line is an imaginary line on the outside of the destruct line. It runs parallel to the destruct line and marks the outer limits of impact for a rocket or missile destroyed under command-destruct procedures.

dogleg   A directional turn made in a launch vehicle's ascent trajectory to produce a more favorable orbit inclination or to avoid passing over a populated (no-fly) region.

double-base propellant   A solid rocket propellant using two unstable compounds, such as nitrocellulose and nitroglycerin. These unstable compounds contain enough chemically bonded oxidizer to sustain combustion.

downlink   The telemetry signal received at a ground station from a spacecraft or space probe.

downrange   A location away from the launch site but along the intended flight path (trajectory) of a missile or rocket flown from a rocket range. For example, the rocket vehicle tracking station on Ascension Island in the South Atlantic Ocean is far downrange from the launch sites at Cape Canaveral Air Force Station in Florida.

dry emplacement   A launch site that has no provision for water cooling of the pad during the launch of a rocket. *Compare with* WET EMPLACEMENT.

Dyna-Soar (Dynamic Soaring)   An early U.S. Air Force space project from 1958 to 1963, involving a crewed boost-glide orbital vehicle that was to be sent into orbit by an expendable launch vehicle, perform its military mission, and return to Earth using wings to glide through the atmosphere during reentry (in a manner similar to NASA's space shuttle). The project was canceled in favor of the civilian (NASA) human space flight program, involving the Mercury Project, Gemini Project, and Apollo Project. Also called the *X-20 Project*.

electric propulsion   A rocket engine that converts electric power into reactive thrust by accelerating an ionized propellant (such as mercury, cesium, argon, or xenon) to a very high exhaust velocity. There are three general types of electric rocket engine: electrothermal, electromagnetic, and electrostatic.

engine cutoff   The specific time when a rocket engine is shut down during a flight.

erosive burning   An increased rate of burning (combustion) that occurs in certain solid propellant rockets as a result of the scouring influence of combustion gases moving at high speed across the burning surface.

escape rocket   A small rocket engine, attached to the leading end of an escape tower, that can provide additional thrust to the crew's space capsule so it can quickly separate from a malfunctioning or exploding expendable launch vehicle during liftoff.

European Space Agency (ESA)   An international organization that promotes the peaceful use of outer space and cooperation among the European member states in space research and applications.

exhaust plume   Hot gas ejected from the thrust chamber of a rocket engine.

exoatmospheric   Occurring outside Earth's atmosphere; events and actions that take place at altitudes above 62 miles (about 100 kilometers).

expendable launch vehicle   A ground-launched rocket vehicle, capable of placing a payload into orbit around Earth or on an Earth-escape trajectory, whose various stages and supporting hardware are not designed or intended for recovery or reuse; a "throwaway" launch vehicle.

experimental vehicle (symbol: X)   A missile, rocket, or aerospace vehicle in the research, development, and testing portion of its technical life cycle; a vehicle not yet approved for operational use.

external tank   The large tank that contains the cryogenic propellants for the three space shuttle main engines. This tank forms the structural backbone of NASA's Space Transportation System flight vehicle.

fallaway section   A section of a rocket vehicle that is cast off and separates from the vehicle during flight, especially a section that falls back to Earth.

ferry flight   An in-the-atmosphere flight of NASA's space shuttle Orbiter vehicle while mated on top of a specially configured Boeing 747 shuttle carrier aircraft.

film cooling   The cooling of body or surface, such as the inner surface of a rocket's combustion chamber, by maintaining a thin fluid layer over the affected area.

fire arrow   An early gunpowder rocket attached to a large bamboo stick; developed by the Chinese about one thousand years ago to confuse and startle enemy troops.

flame bucket   A deep, cavelike metal construction built beneath a launch pad. It is open at the top to receive the hot engine exhaust gases from the rocket positioned above it and has one to three sides open below. During thrust buildup and the beginning of launch, water can be sprayed on the flame bucket to keep it from melting.

flight test vehicle   A rocket, missile, or aerospace vehicle used for performing flight tests that demonstrate the capabilities of the vehicle itself or of specific equipment carried onboard.

force (symbol F)   The cause of the acceleration of material objects as measured by the rate of change of momentum produced on a free body. Force is a vector quantity, mathematically expressed by Sir Isaac Newton's second law of motion: force = mass × acceleration.

free rocket   A rocket not subject to guidance or control in flight.

fuselage   The central part of an aerospace vehicle or aircraft that accommodates crew, passengers, payload, or cargo.

ground support equipment (GSE)   Any nonflight equipment used for launch, checkout, or in-flight support of an expendable rocket, reusable aerospace vehicle, spacecraft, or other type of payload.

guidance system   A system that evaluates flight information; correlates it with target or destination data; determines the desired flight path of the missile, spacecraft, or aerospace vehicle; and communicates the necessary commands to the vehicle's flight control system.

guided missile (GM)   An unmanned, self-propelled vehicle that moves above the surface of Earth whose trajectory or course is capable of being controlled while in flight.

gun-launch-to-space (GLTS)   An advanced launch concept involving the use of a long and powerful electromagnetic launcher to hurl small satellites and payloads into orbit.

gyro   A device that uses the angular momentum of a spinning mass (rotor) to sense angular motion of its base about one or two axes orthogonal (mutually perpendicular) to the spin axis. Also called a gyroscope.

heat soak   The increase in the temperature of rocket engine components after firing has ceased. This occurs because of heat transfer through adjoining parts of the engine when no active cooling is present.

heavy-lift launch vehicle (HLLV)   A conceptual large capacity space-lift vehicle capable of carrying tons of cargo into low Earth orbit at substantially less cost than today's expendable launch vehicles.

hold   To stop the sequence of events during a countdown until an impediment has been removed so that the countdown to launch can be resumed.

holddown test   The test of a rocket, while it is firing but restrained in a test stand.

hot-fire test   A liquid-fuel propulsion system test conducted by actually firing the rocket engine(s) (usually for a short period of time) with the rocket vehicle secured to the launch pad by holddown bolts. *Compare with* COLD-FLOW TEST.

hydrazine (symbol: N_2H_4)   A toxic, colorless liquid that is often used as a rocket propellant because it reacts violently with many oxidizers. It is spontaneously ignitable with concentrated hydrogen peroxide (H_2O_2) and nitric acid.

hydrogen (symbol: H)   A colorless, odorless gas that is the most abundant chemical element in the universe. Hydrogen occurs as molecular hydrogen (H_2), atomic hydrogen (H), and ionized hydrogen (H^+), which is atomic hydrogen broken down into a proton and its companion electron. Hydrogen has three isotopic forms, protium (ordinary hydrogen), deuterium (heavy hydrogen), and tritium (radioactive hydrogen). Liquid hydrogen (LH_2) is an excellent, high-performance cryogenic chemical propellant for rocket engines, especially those using liquid oxygen (LO_2) as the oxidizer.

hypergolic ignition   Ignition that involves no external energy source but results entirely from the spontaneous reaction of two materials (both liquid or a liquid-solid combination) when they are brought into contact.

hypersonic   Pertaining to speeds much greater that the speed of sound, typically speeds of Mach number five (M = 5) and greater.

igniter   A device used to start combustion of a rocket engine.

inertia   The resistance of a body to a change in its state of motion. Mass is an inherent property of a body that helps us quantify inertia. *See also* NEWTON'S LAWS OF MOTION.

inertial upper stage (IUS)   A versatile orbital transfer vehicle developed by the U.S. Air Force that uses solid-propellant rocket motors to boost a payload from low Earth orbit into higher altitude destinations. *See also* UPPER STAGE.

injector   A device that propels (injects) fuel and/or oxidizer into the combustion chamber of a liquid propellant rocket engine. It atomizes and mixes the propellants so they can burn more completely.

insertion   The process of putting an artificial satellite, aerospace vehicle, or spacecraft into orbit.

integration   The collection of activities leading to the compatible assembly of payload and launch vehicle into the desired final (flight) configuration.

intercontinental ballistic missile (ICBM)   A ballistic missile with a range in excess of 3,400 miles (5,500 km).

intermediate range ballistic missile (IRBM)   A ballistic missile with a range capability from about 620 to 3,400 miles (1,000 to 5,500 km).

interplanetary   Between the planets; within the solar system.

interstage section   A section of a missile or rocket that lies between stages. *See also* STAGING.

interstellar   Between or among the stars.

ion   An atom or molecule that has lost or (more rarely) gained one or more electrons. This ionization process electrically charges an atom or molecule.

ion engine   An electrostatic rocket engine in which a propellant (e.g., cesium, mercury, argon, or xenon) is ionized and the propellant ions are accelerated by an imposed electric field to very high exhaust velocity. *See also* ELECTRIC PROPULSION.

jettison   To discard or toss away.

Kapustin Yar   A minor, early Russian launch site that is located on the banks of the Volga River near Volgograd at approximately 48.4 degrees north latitude and 45.8 degrees east longitude.

Kennedy Space Center (KSC)   Sprawling NASA spaceport on the east central coast of Florida adjacent to Cape Canaveral Air Force Station. Launch site (Complex 39) and primary landing/recovery site for the space shuttle.

kinetic energy (common symbols: KE or E_{KE})   The energy an object possesses as a result of its motion. In Newtonian (nonrelativistic) mechanics, kinetic energy is one-half the product of mass (m) and the square of its velocity (v), that is $E_{KE} = 1/2 \, m \, v^2$.

launch   *(noun)* The action that occurs when a rocket or aerospace vehicle propels itself from a planetary surface. *(verb)* To send off a rocket or missile under its own propulsive power.

launch azimuth   The initial compass heading of a rocket vehicle at launch.

launch pad   The load-bearing base or platform from which a rocket, missile, or aerospace vehicle is launched.

launch site   The extensive, well-defined area used to launch rocket vehicles for operational or for test purposes. Also called the launch complex.

launch vehicle (LV)   An expendable or reusable rocket-propelled vehicle that provides sufficient thrust to place a spacecraft in orbit around Earth or to send a payload on an interplanetary trajectory to another celestial body. Sometimes called booster or space lift vehicle.

launch window   An interval of time during which a launch may be made to satisfy some mission objective. Sometimes just a short period each day for a certain number of days.

liftoff   The action of a rocket or aerospace vehicle as it separates from its launch pad in a vertical ascent.

light-year (symbol: ly)   The distance light (or other forms of electromagnetic radiation) can travel in one year. One light-year equals a distance of approximately 5.88×10^{12} miles (9.46×10^{12} km) or 63,240 astronomical units (AU).

liquid hydrogen (LH$_2$)   A cryogenic liquid propellant used as the fuel with liquid oxygen serving as the oxidizer in high performance rocket engines. Hydrogen remains a liquid only at very low (cryogenic) temperatures, typically about 20 K (–253°C; –424°F) or less, imposing special storage and handling requirements.

liquid oxygen (LOX or LO$_2$)   A cryogenic liquid propellant often used as an oxidizer with RP-1 fuel in many expendable launch vehicles or with liquid hydrogen fuel in high performance liquid propellant rocket engines, like NASA's space shuttle main engines. LOX requires storage at temperatures below 90 K (–183°C; –298°F).

liquid propellant   Any combustible liquid fed into the combustion chamber of a liquid-fueled rocket engine.

liquid-propellant rocket engine   A rocket engine that uses chemical propellants in liquid form for both the fuel and oxidizer.

Long March (LM)   A family of expendable launch vehicles developed by China.

low Earth orbit (LEO)   A circular orbit just above Earth's sensible atmosphere at an altitude of between 185 to 250 miles (300 to 400 km).

main stage   For a multistage rocket vehicle, the stage that develops the greatest amount of thrust.

mass (symbol: m)   Mass describes "how much" material makes up an object and gives rise to its inertia. The SI unit for mass is the kilogram (kg). An object that is one pound-mass (0.45 kg) on Earth is also one pound (0.45 kg) of mass on the surface of Mars, or anywhere else in the universe. Remember, an object's mass is *not* its weight. *See also* WEIGHT.

mass fraction   The fraction of a rocket's (or rocket stage's) mass that is taken up by propellant. The remaining mass is structure and payload.

mating   The act of fitting together two major components of an aerospace system, such as the mating of a launch vehicle and its payload—a scientific spacecraft. Also the physical joining of two orbiting spacecraft either through a docking or a berthing process.

meteorological rocket   A sounding rocket designed for routine observation of Earth's upper atmosphere (as opposed to scientific research).

missile   Any object thrown, dropped, fired, launched, or otherwise projected with the purpose of striking a target. Short for ballistic missile or guided missile. "Missile" should *not* be used loosely as an equivalent term for rocket or launch vehicle.

momentum (linear)   The linear momentum (p) of a particle is the product of the particle's mass (m) and its velocity (v). Sir Isaac Newton's second law of motion tells scientists that the time rate of change of momentum of a particle is equal to the resultant force (F) on the particle. *See also* NEWTON'S LAWS OF MOTION.

monopropellant   A liquid propellant for a rocket. It consists of a single chemical substance (such as hydrazine) that decomposes in an exothermic reaction, producing a thrust-generating heated exhaust jet without the use of a second chemical substance.

multistage rocket   A vehicle that has two or more rocket units, each firing after the one behind it has exhausted its propellant. This type of rocket vehicle then discards (or jettisons) each exhausted stage in sequence. Sometimes called a multiple-stage rocket or a step rocket.

NASA   The National Aeronautics and Space Administration, the civilian space agency of the United States. Created in 1958 by an act of Congress, NASA's overall mission is to plan, direct, and conduct civilian (including scientific) aeronautical and space activities for peaceful purposes.

newton (symbol: N)   The SI unit of force, named after Sir Isaac Newton (1642–1727). One newton is the amount of force that gives a one kilogram mass an acceleration of one meter per second per second.

Newton's law of gravitation   The physical law proposed by Sir Isaac Newton (1642–1727) in about 1687. This law states that every particle of matter in the universe attracts every other particle with the force of gravitational attraction (F_G) acting along the line joining the two particles and being proportional to the product of the particle masses (m_1 and m_2), and inversely proportional to the square of the distance (r) between the particles. This law expressed as an equation is $F_G = [G\, m_1 m_2] / r^2$, where G is the universal gravitational constant (approximately 3.44×10^{-8} lb-force-$ft^2/slug^2$ [or 6.6732×10^{-11} N m^2/kg^2 in SI units]).

Newton's laws of motion   The three postulates of motion formulated by Sir Isaac Newton (1642–1727) in about 1685. His first law (the conservation of momentum) states that a body continues in a state of uniform motion (or rest) unless acted upon by an external force. The second law states that the rate of change of momentum of a body is proportional to the force acting upon the body and occurs in the direction of the applied force. The third law (the action and reaction principle) states that for every force acting upon a body, there is a corresponding force of the same magnitude exerted by the body in the opposite direction. The third law is the basic principle by which every rocket operates.

nose cone   The cone-shaped leading edge of a rocket vehicle, which contains and protects the payload or warhead.

nozzle   A flow device that promotes the efficient expansion of the hot gases from the combustion chamber of a rocket. As these gases leave at high velocity, a propulsive (forward) thrust also occurs in accordance with Newton's third law of motion (action-reaction principle).

nuclear electric propulsion (NEP)   A space-deployed propulsion system that uses a space-qualified, compact nuclear reactor to produce the electricity needed to operate a space vehicle's electric propulsion engine(s).

nuclear thermal rocket (NTR)   A rocket vehicle that derives its propulsive thrust from nuclear energy. It uses a nuclear reactor to heat hydrogen to extremely high temperatures before expelling it through a thrust-producing nozzle. *See also* NUCLEAR ELECTRIC PROPULSION.

ogive   The tapered or curved front of a missile or rocket.

orbital transfer vehicle (OTV)   A propulsion system used to transfer a payload from one orbital location to another. An expendable (one-shot) orbital transfer vehicle is often called an upper-stage unit, while a reusable OTV is called a space tug.

orbiter (space shuttle)   The winged aerospace vehicle portion of NASA's space shuttle. It carries astronauts and payload into orbit and returns from outer space by gliding and landing like an airplane. In 2005, NASA's operational orbiter vehicle (OV) fleet included: *Discovery* (OV-103), *Atlantis* (OV-104), and *Endeavour* (OV-105).

outer space   Any region beyond Earth's atmospheric envelope—generally considered to begin at between 62 miles and 124 miles (100 and 200 km) altitude.

oxidizer   A substance whose main function is to supply oxygen for the burning of a rocket engine's fuel.

pad   The platform from which a rocket vehicle is launched. *See also* LAUNCH PAD.

payload   That which a rocket, aerospace vehicle, or spacecraft carries over and above what is necessary for the operation of the vehicle during flight.

payload assist module (PAM)   A family of commercially developed upper-stage vehicles for use with NASA's space shuttle or with expendable launch vehicles, such as the Delta.

payload bay   The large and long enclosed volume within NASA's space shuttle orbiter vehicle designed to carry a wide variety of payloads, including upper-stage vehicles, deployable spacecraft, and attached equipment. Also called cargo bay.

Pegasus (space launch vehicle)   An aircraft-launched space booster rocket capable of placing small payloads and spacecraft into low Earth orbit.

pitch   The rotation (angular motion) of an aerospace vehicle, rocket, or spacecraft about its lateral axis. *See also* ROLL; YAW.

pitchover   The programmed turn from the vertical that a launch vehicle (under power) takes as it describes an arc and points in a direction other than vertical.

plasma   An electrically neutral gaseous mixture of positive and negative ions; sometimes called the fourth state of matter, because it behaves quite differently from solids, liquids, or gases.

Plesetsk   The northern Russian launch site about 185 miles (300 km) south of Archangel that supports a wide variety of military space launches, ballistic missile testing, and scientific spacecraft requiring a polar orbit.

plume   The hot, bright exhaust gases from a rocket.

posigrade rocket   An auxiliary rocket that fires in the direction in which the vehicle is pointed; often used in separating two stages of a launch vehicle or a payload from the upper-stage propulsion system. The firing of a posigrade rocket adds to a spacecraft's speed, while the firing of a retrograde rocket slows it down.

potential energy (symbol: PE)   Energy or ability to do work possessed by an object by virtue of its position in a gravity field above some reference position or datum.

power (symbol: P)   The rate at which work is done or at which energy is transformed per unit time. The British thermal unit per hour (Btu/hr) is the fundamental unit of power in the American engineering system of units; the watt (W) is the fundamental SI unit of power.

progressive burning   A design condition for a solid-propellant rocket in which the surface area of burning propellant increases with time, thereby increasing thrust for some specific period of operation.

propellant   The material, such as a chemical fuel and oxidizer combination, carried in a rocket vehicle, "energized," and then ejected at high velocity as thrust-producing reaction mass. Chemical propellants are either liquid or solid form. Modern launch vehicles use one of three general types of liquid propellants: petroleum-based, cryogenic (very cold), or hypergolic (self-igniting upon contact).

propulsion system   The launch vehicle (or space vehicle) system that includes rocket engines, propellant tanks, fluid lines, and all associated equipment necessary to provide the propulsive force (thrust) as specified for the vehicle.

Proton   A Russian liquid-propellant expendable launch vehicle capable of placing a 46,200 pound-mass (21,000-kg) payload into low Earth orbit (LEO) and sending spacecraft on interplanetary trajectories.

purge   The process of removing residual fuel or oxidizer from the tanks or lines of a liquid-propellant rocket after a test firing.

radial-burning   A solid propellant rocket grain that burns in the radial direction, either outwardly (called an internally burning grain) or inwardly.

reaction engine   An engine that develops thrust by its physical reaction to the ejection of a substance from it. Normally, a reaction engine ejects a stream of hot gases created by internally combusting a propellant—but more advanced reaction engine concepts involve the ejection of photons or nuclear particles. Both rocket and jet propulsion systems involve reaction engines that make use of Sir Isaac Newton's third law of motion (the action-reaction principle).

reentry   The return of objects, originally launched from Earth, back into the sensible atmosphere; the action involved in this event. The major types of reentry are: ballistic, gliding, and skip. When a piece of space debris undergoes an uncontrolled ballistic reentry, it usually burns up in the atmosphere due to excessive aerodynamic heating. An aerospace vehicle, like NASA's space shuttle, makes a controlled atmospheric reentry by using a gliding trajectory designed to carefully dissipate its kinetic energy and potential energy prior to landing.

reentry vehicle (RV)   The part of a rocket or space vehicle designed to reenter Earth's atmosphere in the terminal portion of its trajectory; for example, the nose-cone portion of an intercontinental ballistic missile is designed to survive the aerodynamic heating of reentry and to protect its payload (a nuclear warhead) while it descends on a ballistic trajectory to its target.

regressive burning   For a solid propellant rocket the condition in which the burning surface of the propellant decreases with time—thereby decreasing the pressure and thrust.

retrorocket (retrograde rocket)   An auxiliary rocket that fires in the direction opposite to which a space vehicle is traveling (pointed); low-thrust retrograde rockets produce a retarding force that opposes the vehicle's forward motion and reduces its velocity. *Compare with* POSIGRADE ROCKET.

reusable launch vehicle (RLV)   A conceptual space launch vehicle that includes simple, fully reusable designs which support flexible "airline-type" operations and greatly reduced costs per kilogram of payload delivered into low Earth orbit. These design goals would primarily be achieved through the use of advanced space technology and innovative operational techniques. In 2001, NASA canceled the X-33 prototype RLV program, citing technical difficulties and cost overruns.

rocket   A completely self-contained projectile or flying vehicle propelled by a reaction engine. Since a rocket carries all of its required propellant, it can function in the vacuum of outer space and represents the key to space travel. This fact was independently recognized early in the 20th century by the founders of astronautics: Konstantin Tsiolkovsky, Robert Goddard, and Hermann Oberth. Rockets obey Sir Isaac Newton's third law of motion (the action/reaction principle). There are chemical rockets, nuclear rockets, and electric propulsion rockets. Chemical rockets are further divided into solid-propellant rockets and liquid-propellant rockets.

roll   The rotational or oscillatory movement of an aerospace vehicle or rocket about its longitudinal (lengthwise) axis. *See also* PITCH; YAW.

Rover Program   The overall U.S. nuclear rocket development program from 1959 to 1973.

RP-1   Rocket Propellant Number One (RP-1)—a commonly used hydrocarbon-based, liquid-propellant rocket engine fuel that is refined kerosene.

rumble   A form of combustion instability in a liquid-propellant rocket engine, characterized by a low-pitched, low-frequency rumbling noise.

Saturn (launch vehicle)   Family of powerful expendable launch vehicles developed for NASA by Wernher Von Braun (1912–77) to carry astronauts to the Moon in the Apollo Project.

Scout   A four-stage, solid propellant rocket developed by NASA and used as an expendable launch vehicle to place small (typically 440 pound-mass [200-kg] or less) payloads into low Earth orbit or on suborbital trajectories.

screaming   For a liquid-propellant rocket engine, a relatively high-frequency form of combustion instability, characterized by a high-pitched noise.

scrub   To cancel or postpone a rocket firing, either before or during the countdown.

sloshing   The back-and-forth movement of a liquid propellant in its tank(s), which creates problems of stability and control problems for the rocket vehicle. Engineers often use anti-slosh baffles in the propellant tanks to avoid this problem.

slug   A contrived mass unit in the American engineering unit system equivalent to 32.174 pounds-mass. From Sir Isaac Newton's second law of motion, one pound-force will accelerate a one slug mass at a rate of 1 ft/s^2.

solar electric propulsion (SEP)   A low-thrust propulsion system that uses solar cells to provide the electricity for a spacecraft's electric propulsion rocket engines.

solid propellant rocket   A rocket propelled by a chemical mixture of fuel and oxidizer in solid form and intimately mixed into a monolithic (but not powdered) grain. Sometimes called a solid rocket.

solid rocket booster (SRB)   The two very large solid propellant rockets that operate in parallel to augment the thrust of the Space Shuttle's three main engines for the first two minutes after launch. Each SRB develops about 2.65 million pounds-force (11.8 meganewtons) of thrust at liftoff. After burning for about 120 seconds, the depleted SRBs are jettisoned from the space shuttle and recovered in the Atlantic Ocean downrange of Cape Canaveral for refurbishment and propellant reloading.

sounding rocket   A solid propellant rocket used to carry scientific instruments on parabolic trajectories into the upper regions of Earth's sensible atmosphere and into near-Earth space.

Soyuz (launch vehicle)   The "workhorse" Soviet (and later Russian) launch vehicle that was first used in 1963. With its two cryogenic stages and four cryogenic strap-on engines, this vehicle is capable of placing payloads of up to 15,180 pounds-mass (6,900 kg) into low Earth orbit. At present, it is the most frequently flown launch vehicle in the world. Since 1964, the Soyuz rocket has also been used to launch every Russian human crew space mission.

space launch vehicle (SLV)   The expendable or reusable rocket-propelled vehicle(s) used to lift a payload or spacecraft from the surface of Earth and place it in orbit around the planet or on an interplanetary trajectory.

spaceport   A facility that serves as both a doorway to outer space from the surface of a planet and a port of entry for aerospace vehicles returning from space to the planet's surface. NASA's Kennedy Space Center with its space shuttle launch site and landing complex is an example.

spaceship   An interplanetary spacecraft that carries a human crew.

space shuttle   The major space flight component NASA's Space Transportation System consisting of a winged orbiter vehicle, three space shuttle main engines, the giant external tank, which feeds liquid hydrogen and liquid oxygen to the shuttle's three main liquid propellant rocket engines, and the two solid rocket boosters.

Space Transportation System (STS)   The official name for NASA's space shuttle.

space vehicle   The general term describing a crewed or robot vehicle capable of traveling through outer space. An aerospace vehicle can operate both in outer space and in Earth's atmosphere.

specific impulse (I_{sp})   An index of performance for rocket engines and their various propellant combinations—defined as the thrust produced by propellant combustion divided by the propellant mass flow rate.

speed of light (symbol: c)   The speed at which electromagnetic radiation (including light) moves through a vacuum; a universal constant equal to approximately 186,000 miles per second (300,000 km/s).

spin stabilization   Directional stability of a missile or spacecraft obtained as a result of spinning the moving body about its axis of symmetry.

staging   The practice of placing smaller rockets on top of larger ones, thereby increasing the ability of the combination to lift larger payloads or to give a particular payload a higher final velocity. In multistage rockets, the stages are numbered chronologically in the order of burning (i.e., first stage, second stage, third stage, etc.). When the first stage stops burning, it separates from the rest of the vehicle and falls away. Then the second stage rocket ignites, fires until burnout, and also separates. The staging process continues up to the last stage, which contains the payload. In this way, the mass of empty propellant tanks is discarded during the ascent.

starship   A conceptual, very advanced space vehicle capable of traveling the great distances between star systems within decades or less. The term "starship" is generally reserved for vehicles that could carry intelligent beings, while interstellar probe applies to an advanced robot spacecraft capable of interstellar travel at ten percent or more of the speed of light.

storable propellant   Rocket propellant (usually liquid) that can be stored for prolonged periods without special temperature or pressure environments.

Thor   An intermediate-range ballistic missile developed by the U.S. Air Force in the late 1950s; also used by NASA and the military as a space launch vehicle.

throttling   The variation of the thrust of a rocket engine during powered flight.

thrust (symbol: T)   The forward force provided by a reaction engine, such as a rocket.

Titan (launch vehicle)   The family of powerful U.S. Air Force ballistic missiles and expendable launch vehicles that began in 1955 with the Titan I—the first American two-stage intercontinental ballistic missile. The Titan IV is the newest and most powerful member.

trajectory   The three-dimensional path traced by any object moving because of an externally applied force; the flight path of a space vehicle.

turbopump system   The high-speed pumping equipment in a liquid propellant rocket engine, designed to raise the pressure of the propellants (fuel and oxidizer) so they can go from the tanks into the combustion chamber at specified flow rates.

ullage   The amount that a container, such as a propellant tank, lacks of being full.

umbilical   An electrical or fluid servicing line between the ground or tower and an upright rocket vehicle before launch.

upper stage   The second, third, or later rocket stage of a multistage rocket vehicle. Once lifted into low Earth orbit, a spacecraft often uses an attached upper stage to reach its final destination—a higher altitude orbit around Earth or an interplanetary trajectory.

Vandenberg Air Force Base   Located on the central California coast north of Santa Barbara, this U.S. Air Force facility is the launch site of all military, NASA, and commercial space launches that require high inclination, especially polar orbits.

vector   A physical quantity, such as force, velocity, or acceleration that has both magnitude and direction at each point in space, as opposed to a scalar quantity which just has magnitude.

vengeance weapon 2 (V-2)   The V-2 or "Vergeltungwaffe 2" was the first modern military ballistic missile. This liquid propellant rocket was designed and flown by the German Army during World War II and then became the technical ancestor for many large American and Russian rockets constructed during the cold war.

vernier engine   A rocket engine of small thrust used primarily to obtain a fine adjustment in the velocity and trajectory or in the attitude of a rocket or aerospace vehicle.

warhead   The payload of a ballistic missile or military rocket, usually a nuclear weapon or chemical high explosive.

weight (symbol w)   The product of the mass (m) of an object on the surface of a planetary body and the gravitational acceleration (g) acting on the object—that is, $w = mg$. The weight is a force, expressed in units of pounds-force (newtons). For example, a one pound-mass (0.45-kg) object has a weight of one-pound-force (or 4.448 newtons) on Earth's surface.

wet emplacement   A launch pad that provides a deluge of water for cooling the flame bucket and other equipment during a rocket firing.

yaw   The rotation or oscillation of a missile or aerospace vehicle about its vertical axis so as to cause the longitudinal axis of the vehicle to deviate from the flight line or heading in its horizontal plane. *See also* PITCH; ROLL.

Zenith   A three-stage Russian expendable launch vehicle capable of placing payloads up to about 30,800 pounds-mass (14,000 kg) into low Earth orbit.

Further Reading

· ·

RECOMMENDED BOOKS

Angelo, Joseph A., Jr. *The Dictionary of Space Technology.* Rev. ed. New York: Facts On File, Inc., 2004.

————. *Encyclopedia of Space Exploration.* New York: Facts On File, Inc., 2000.

————, and Irving W. Ginsberg, eds. *Earth Observations and Global Change Decision Making, 1989: A National Partnership.* Malabar, Fla.: Krieger Publishing, 1990.

Brown, Robert A., ed. *Endeavour Views the Earth.* New York: Cambridge University Press, 1996.

Burrows, William E., and Walter Cronkite. *The Infinite Journey: Eyewitness Accounts of NASA and the Age of Space.* Discovery Books, 2000.

Chaisson, Eric, and Steve McMillian. *Astronomy Today.* 5th ed. Upper Saddle River, N.J.: Pearson Prentice Hall, 2005.

Cole, Michael D. *International Space Station. A Space Mission.* Springfield, N.J.: Enslow Publishers, 1999.

Collins, Michael. *Carrying the Fire.* New York: Cooper Square Publishers, 2001.

Consolmagno, Guy J., et al. *Turn Left at Orion: A Hundred Night Objects to See in a Small Telescope—And How to Find Them.* New York: Cambridge University Press, 2000.

Damon, Thomas D. *Introduction to Space: The Science of Spaceflight.* 3d ed. Malabar, Fla.: Krieger Publishing Co., 2000.

Dickinson, Terence. *The Universe and Beyond.* 3d ed. Willowdater, Ont.: Firefly Books Ltd., 1999.

Heppenheimer, Thomas A. *Countdown: A History of Space Flight.* New York: John Wiley and Sons, 1997.

Kluger, Jeffrey. *Journey beyond Selene: Remarkable Expeditions Past Our Moon and to the Ends of the Solar System.* New York: Simon & Schuster, 1999.

Kraemer, Robert S. *Beyond the Moon: A Golden Age of Planetary Exploration, 1971–1978.* Smithsonian History of Aviation and Spaceflight Series. Washington, D.C.: Smithsonian Institution Press, 2000.

Lewis, John S. *Rain of Iron and Ice: The Very Real Threat of Comet and Asteroid Bombardment.* Reading, Mass.: Addison-Wesley, 1996.

Logsdon, John M. *Together in Orbit: The Origins of International Participation in the Space Station.* NASA History Division, Monographs in Aerospace History 11, Washington, D.C.: Office of Policy and Plans, November 1998.

Matloff, Gregory L. *The Urban Astronomer: A Practical Guide for Observers in Cities and Suburbs.* New York: John Wiley and Sons, 1991.

Neal, Valerie, Cathleen S. Lewis, and Frank H. Winter. *Spaceflight: A Smithsonian Guide.* New York: Macmillan, 1995.

Pebbles, Curtis L. *The Corona Project: America's First Spy Satellites.* Annapolis, Md.: Naval Institute Press, 1997.

Seeds, Michael A. Horizons: *Exploring the Universe.* 6th ed. Pacific Grove, Calif.: Brooks/Cole Publishing, 1999.

Sutton, George Paul. *Rocket Propulsion Elements.* 7th ed. New York: John Wiley & Sons, 2000.

Todd, Deborah, and Joseph A. Angelo, Jr. *A to Z of Scientists in Space and Astronomy.* New York: Facts On File, Inc., 2005.

EXPLORING CYBERSPACE

In recent years, numerous Web sites dealing with astronomy, astrophysics, cosmology, space exploration, and the search for life beyond Earth have appeared on the Internet. Visits to such sites can provide information about the status of ongoing missions, such as NASA's *Cassini* spacecraft as it explores the Saturn system. This book can serve as an important companion, as you explore a new Web site and encounter a person, technology phrase, or physical concept unfamiliar to you and not fully discussed within the particular site. To help enrich the content of this book and to make your astronomy and/or space technology–related travels in cyberspace more enjoyable and productive, the following is a selected list of Web sites that are recommended for your viewing. From these sites you will be able to link to many other astronomy or space-related locations on the Internet. Please note that this is obviously just a partial list of the many astronomy and space-related Web sites now available. Every effort has been made at the time of publication to ensure the accuracy of the information provided. However, due to the dynamic nature of the Internet, URL changes do occur and any inconvenience you might experience is regretted.

Selected Organizational Home Pages

European Space Agency (ESA) is an international organization whose task is to provide for and promote, exclusively for peaceful purposes, cooperation among European states in space research and technology and their applications. URL: http://www.esrin.esa.it. Accessed on April 12, 2005.

National Aeronautics and Space Administration (NASA) is the civilian space agency of the United States government and was created in 1958 by an act of Congress. NASA's overall mission is to plan, direct, and conduct American civilian (including scientific) aeronautical and space activities for peaceful purposes. URL: http://www.nasa.gov. Accessed on April 12, 2005.

National Oceanic and Atmospheric Administration (NOAA) was established in 1970 as an agency within the U.S. Department of Commerce to ensure the safety of the general public from atmospheric phenomena and to provide the public with an understanding of Earth's environment and resources. URL: http://www.noaa.gov. Accessed on April 12, 2005.

National Reconnaissance Office (NRO) is the organization within the Department of Defense that designs, builds, and operates U.S. reconnaissance satellites. URL: http://www.nro.gov. Accessed on April 12, 2005.

United States Air Force (USAF) serves as the primary agent for the space defense needs of the United States. All military satellites are launched from Cape Canaveral Air Force Station, Florida or Vandenberg Air Force Base, California. URL: http://www.af.mil. Accessed on April 14, 2005.

United States Strategic Command (USSTRATCOM) is the strategic forces organization within the Department of Defense, which commands and controls U.S. nuclear forces and military space operations. URL: http://www.stratcom. mil. Accessed on April 14, 2005.

Selected NASA Centers

Ames Research Center (ARC) in Mountain View, California, is NASA's primary center for exobiology, information technology, and aeronautics. URL: http://www.arc.nasa.gov. Accessed on April 12, 2005.

Dryden Flight Research Center (DFRC) in Edwards, California, is NASA's center for atmospheric flight operations and aeronautical flight research. URL: http://www.dfrc.nasa.gov. Accessed on April 12, 2005.

Glenn Research Center (GRC) in Cleveland, Ohio, develops aerospace propulsion, power, and communications technology for NASA. URL: http://www.grc.nasa.gov. Accessed on April 12, 2005.

Goddard Space Flight Center (GSFC) in Greenbelt, Maryland, has a diverse range of responsibilities within NASA, including Earth system science, astrophysics, and operation of the *Hubble Space Telescope* and other Earth-orbiting spacecraft. URL: http://www.nasa.gov/goddard. Accessed on April 14, 2005.

Jet Propulsion Laboratory (JPL) in Pasadena, California, is a government-owned facility operated for NASA by Caltech. JPL manages and operates NASA's deep-space scientific missions, as well as the NASA's Deep Space Network, which communicates with solar system exploration spacecraft. URL: http://www.jpl.nasa.gov. Accessed on April 12, 2005.

Johnson Space Center (JSC) in Houston, Texas, is NASA's primary center for design, development, and testing of spacecraft and associated systems for human space flight, including astronaut selection and training. URL: http://www.jsc.nasa.gov. Accessed on April 12, 2005.

Kennedy Space Center (KSC) in Florida is the NASA center responsible for ground turnaround and support operations, prelaunch checkout, and launch of the space shuttle. This center is also responsible for NASA launch facilities at Vandenberg Air Force Base, California. URL: http://www.ksc.nasa.gov. Accessed on April 12, 2005.

Langley Research Center (LaRC) in Hampton, Virginia, is NASA's center for structures and materials, as well as hypersonic flight research and aircraft safety. URL: http://www.larc.nasa.gov. Accessed on April 15, 2005.

Marshall Space Flight Center (MSFC) in Huntsville, Alabama, serves as NASA's main research center for space propulsion, including contemporary rocket engine development as well as advanced space transportation system concepts. URL: http://www.msfc.nasa.gov. Accessed on April 12, 2005.

Stennis Space Center (SSC) in Mississippi is the main NASA center for large rocket engine testing, including space shuttle engines as well as future generations of space launch vehicles. URL: http://www.ssc.nasa.gov. Accessed on April 14, 2005.

Wallops Flight Facility (WFF) in Wallops Island, Virginia, manages NASA's suborbital sounding rocket program and scientific balloon flights to Earth's upper atmosphere. URL: http://www.wff.nasa.gov. Accessed on April 14, 2005.

White Sands Test Facility (WSTF) in White Sands, New Mexico, supports the space shuttle and space station programs by performing tests on and evaluating potentially hazardous materials, space flight components, and rocket propulsion systems. URL: http://www.wstf.nasa.gov. Accessed on April 12, 2005.

Selected Space Missions

Cassini Mission is an ongoing scientific exploration of the planet Saturn. URL: http://saturn.jpl.nasa.gov. Accessed on April 14, 2005.

Chandra X-ray Observatory (CXO) is a space-based astronomical observatory that is part of NASA's Great Observatories Program. *CXO* observes the universe in the X-ray portion of the electromagnetic spectrum. URL: http://www.chandra.harvard.edu. Accessed on April 14, 2005.

Exploration of Mars is the focus of this Web site, which features the results of numerous contemporary and previous flyby, orbiter, and lander robotic spacecraft. URL: http://mars.jpl.nasa.gov. Accessed on April 14, 2005.

National Space Science Data Center (NSSDC) provides a worldwide compilation of space missions and scientific spacecraft. URL: http://nssdc.gsfc.nasa.gov/planetary. Accessed on April 14, 2005.

Voyager (Deep Space/Interstellar) updates the status of NASA's *Voyager 1* and *2* spacecraft as they travel beyond the solar system. URL: http://voyager.jpl.nasa.gov. Accessed on April 14, 2005.

Other Interesting Astronomy and Space Sites

Arecibo Observatory in the tropical jungle of Puerto Rico is the world's largest radio/radar telescope. URL: http://www.naic.edu. Accessed on April 14, 2005.

Astrogeology (USGS) describes the USGS Astrogeology Research Program, which has a rich history of participation in space exploration efforts and planetary mapping. URL: http://planetarynames.wr.usgs.gov. Accessed on April 14, 2005.

Hubble Space Telescope **(HST)** is an orbiting NASA Great Observatory that is studying the universe primarily in the visible portions of the electromagnetic spectrum. URL: http://hubblesite.org. Accessed on April 14, 2005.

NASA's Deep Space Network (DSN) is a global network of antennas that provide telecommunications support to distant interplanetary spacecraft and probes. URL: http://deepspace.jpl.nasa.gov/dsn. Accessed on April 14, 2005.

NASA's Space Science News provides contemporary information about ongoing space science activities. URL: http://science.nasa.gov. Accessed on April 14, 2005.

National Air and Space Museum (NASM) of the Smithsonian Institution in Washington, D.C., maintains the largest collection of historic aircraft and spacecraft in the world. URL: http://www.nasm.si.edu. Accessed on April 14, 2005.

Planetary Photojournal is a NASA/JPL– sponsored Web site that provides an extensive collection of images of celestial objects within and beyond the solar system, historic and contemporary spacecraft used in space exploration, and advanced aerospace technologies URL: http://photojournal.jpl.nasa.gov. Accessed on April 14, 2005.

Planetary Society is the nonprofit organization founded in 1980 by Carl Sagan and other scientists that encourages all spacefaring nations to explore other worlds. URL: http://planetary.org. Accessed on April 14, 2005.

Search for Extraterrestrial Intelligence (SETI) Projects at UC Berkeley is a Web site that involves contemporary activities in the search for extraterrestrial intelligence (SETI), especially a radio SETI project that lets anyone with a computer and an Internet connection participate. URL: http://www.setiathome.ssl.berkeley.edu. Accessed on April 14, 2005.

Solar System Exploration is a NASA-sponsored and -maintained Web site that presents the last events, discoveries and missions involving the exploration of the solar system. URL: http://solarsystem.nasa.gov. Accessed on April 14, 2005.

Space Flight History is a gateway Web site sponsored and maintained by the NASA Johnson Space Center. It provides access to a wide variety of interesting data and historic reports dealing with (primarily U.S.) human space flight. URL: http://www11.jsc.nasa.gov/history. Accessed on April 14, 2005.

Space Flight Information (NASA) is a NASA-maintained and -sponsored gateway Web site that provides the latest information about human spaceflight activities, including the *International Space Station* and the space shuttle. URL: http://spaceflight.nasa.gov Accessed on April 14, 2005.

Index